Practical Time Series Analysis

Master Time Series Data Processing, Visualization, and
Modeling using Python

Dr. Avishek Pal
Dr. PKS Prakash

BIRMINGHAM - MUMBAI

Practical Time Series Analysis

First published: September 2017

Production reference: 2041017

Published by Packt Publishing Ltd.
Livery Place
35 Livery Street
Birmingham
B3 2PB, UK.
ISBN 978-1-78829-022-7

www.packtpub.com

Credits

Authors
Dr. Avishek Pal
Dr. PKS Prakash

Reviewer
Prabhanjan Tattar

Commissioning Editor
Veena Pagare

Acquisition Editor
Aman Singh

Content Development Editor
Snehal Kolte

Technical Editor
Danish Shaikh

Copy Editor
Tasneem Fatehi

Project Coordinator
Manthan Patel

Proofreader
Safis Editing

Indexer
Tejal Daruwale Soni

Graphics
Tania Dutta

Production Coordinator
Deepika Naik

About the Authors

Dr. Avishek Pal, PhD, is a software engineer, data scientist, author, and an avid Kaggler living in Hyderabad, the City of Nawabs, India. He has a bachelor of technology degree in industrial engineering from the Indian Institute of Technology (IIT) Kharagpur and has earned his doctorate in 2015 from University of Warwick, Coventry, United Kingdom. At Warwick, he studied at the prestigious Warwick Manufacturing Centre, which functions as one of the centers of excellence in manufacturing and industrial engineering research and teaching in UK.

In terms of work experience, Avishek has a diversified background. He started his career as a software engineer at IBM India to develop middleware solutions for telecom clients. This was followed by stints at a start-up product development company followed by Ericsson, a global telecom giant. During these three years, Avishek lived his passion for developing software solutions for industrial problems using Java and different database technologies.

Avishek always had an inclination for research and decided to pursue his doctorate after spending three years in software development. Back in 2011, the time was perfect as the analytics industry was getting bigger and data science was emerging as a profession. Warwick gave Avishek ample time to build up the knowledge and hands-on practice on statistical modeling and machine learning. He applied these not only in doctoral research, but also found a passion for solving data science problems on Kaggle.

After doctoral studies, Avishek started his career in India as a lead machine learning engineer for a leading US-based investment company. He is currently working at Microsoft as a senior data scientist and enjoys applying machine learning to generate revenue and save costs for the software giant.

Avishek has published several research papers in reputed international conferences and journals. Reflecting back on his career, he feels that starting as a software developer and then transforming into a data scientist gives him the end-to-end focus of developing statistics into consumable software solutions for industrial stakeholders.

I would like to thank my wife for putting up with my late-night writing sessions and weekends when I had to work on this book instead of going out. Thanks also goes to Prakash, the co-author of this book, for encouraging me to write a book.

I would also like to thank my mentors with whom I have interacted over the years. People such as Prof. Manoj Kumar Tiwari from IIT Kharagpur and Prof. Darek Ceglarek, my doctoral advisor at Warwick, have taught me and showed me the right things to do, both academically and career-wise.

Dr. PKS Prakash is a data scientist and author. He has spent the last 12 years in developing many data science solutions in several practice areas within the domains of healthcare, manufacturing, pharmaceutical, and e-commerce. He is working as the data science manager at ZS Associates. ZS is one of the world's largest business services firms, helping clients with commercial success by creating data-driven strategies using advanced analytics that they can implement within their sales and marketing operations in order to make them more competitive, and by helping them deliver an impact where it matters.

Prakash's background involves a PhD in industrial and system engineering from Wisconsin-Madison, US. He has earned his second PhD in engineering from University of Warwick, UK. His other educational qualifications involve a masters from University of Wisconsin-Madison, US, and bachelors from National Institute of Foundry and Forge Technology (NIFFT), India. He is the co-founder of Warwick Analytics spin-off from University of Warwick, UK.

Prakash has published articles widely in research areas of operational research and management, soft computing tools, and advance algorithms in leading journals such as IEEE-Trans, EJOR, and IJPR among others. He has edited an issue on *Intelligent Approaches to Complex Systems* and contributed in books such as *Evolutionary Computing in Advanced Manufacturing* published by WILEY and *Algorithms and Data Structures using R* and *R Deep Learning Cookbook* published by PACKT.

I would like to thank my wife, Dr. Ritika Singh, and daughter, Nishidha Singh, for all their love and support. I would also like to thank Aman Singh (Acquisition Editor) of this book and the entire PACKT team whose names may not all be enumerated but their contribution is sincerely appreciated and gratefully acknowledged.

About the Reviewer

Prabhanjan Tattar is currently working as a Senior Data Scientist at Fractal Analytics Inc. He has 8 years of experience as a statistical analyst. Survival analysis and statistical inference are his main areas of research/interest, and he has published several research papers in peer-reviewed journals and also authored two books on R: R Statistical Application Development by Example, Packt Publishing, and A Course in Statistics with R, Wiley. The R packages, gpk, RSADBE, and ACSWR are also maintained by him.

www.PacktPub.com

For support files and downloads related to your book, please visit www.PacktPub.com.

Did you know that Packt offers eBook versions of every book published, with PDF and ePub files available? You can upgrade to the eBook version at www.PacktPub.com and as a print book customer, you are entitled to a discount on the eBook copy. Get in touch with us at service@packtpub.com for more details.

At www.PacktPub.com, you can also read a collection of free technical articles, sign up for a range of free newsletters and receive exclusive discounts and offers on Packt books and eBooks.

https://www.packtpub.com/mapt

Get the most in-demand software skills with Mapt. Mapt gives you full access to all Packt books and video courses, as well as industry-leading tools to help you plan your personal development and advance your career.

Why subscribe?

- Fully searchable across every book published by Packt
- Copy and paste, print, and bookmark content
- On demand and accessible via a web browser

Customer Feedback

Thanks for purchasing this Packt book. At Packt, quality is at the heart of our editorial process. To help us improve, please leave us an honest review on this book's Amazon page at `https://www.amazon.com/dp/1788290224`.

If you'd like to join our team of regular reviewers, you can e-mail us at `customerreviews@packtpub.com`. We award our regular reviewers with free eBooks and videos in exchange for their valuable feedback. Help us be relentless in improving our products!

Table of Contents

Preface

This book is about an introduction to time series analysis using Python. We aim to give you a clear overview of the basic concepts of the discipline and describe useful techniques that would be applicable for commonly-found analytics use cases in the industry. With too many projects requiring trend analytics and forecasting based on past data, time series analysis is an important tool in the knowledge arsenal of any modern data scientist. This book will equip you with tools and techniques, which will let you confidently think through a problem and come up with its solution in time series forecasting.

Why Python? Python is rapidly becoming a first choice for data science projects across different industry sectors. Most state-of-the art machine learning and deep learning libraries have a Python API. As a result, many data scientists prefer Python to implement the entire project pipeline that consists of data wrangling, model building, and model validation. Besides, Python provides easy-to-use APIs to process, model, and visualize time series data. Additionally, Python has been a popular language for the development of backend for web applications and hence has an appeal to a wider base of software professionals.

Now, let's see what you can expect to learn from every chapter this book.

What this book covers

Chapter 1, *Introduction to Time Series*, starts with a discussion of the three different types of datasets—cross-section, time series, and panel. The transition from cross-sectional to time series and the added complexity of data analysis is discussed. Special mathematical properties that make time series data special are described. Several examples demonstrate how exploratory data analysis can be used to visualize these properties.

Chapter 2, *Understanding Time Series Data*, covers three topics, advanced preprocessing and visualization of time series data through *resampling, group-by,* and calculation of moving averages; stationarity and statistical hypothesis testing to detect stationarity in a time series; and various methods of time series decomposition for stationarizing a non-stationary time series.

Chapter 3, *Exponential Smoothing based Methods*, covers smoothing-based models using the Holt-Winters approach for first order to capture levels, second order to smoothen levels and trend, and higher order smoothing is illustrated, which captures level, trend, and seasonality within a time series dataset.

Chapter 4, *Auto-Regressive Models*, discusses autoregressive models for forecasting. The chapter covers a detailed implementation for moving average (MA), autoregressive (AR), Auto Regressive Moving Average (ARMA), and Auto Regressive Integrated Moving Average (ARIMA) to capture different levels of nuisance within time series data during forecasting.

Chapter 5, *Deep Learning for Time Series Forecasting*, discusses recent deep learning algorithms that can be directly adapted to develop forecasting models for time series data. Recurrent Neural Networks (RNNs) are a natural choice for modeling sequence in data. In this chapter, different RNNs such as Vanilla RNN, Gated Recurrent Units, and Long Short Term Memory units are described to develop forecasting models on time series data. The mathematical formulations involved in developing these RNNs are conceptually discussed. Case studies are solved using the 'keras' deep learning library of Python.

Appendix, *Getting Started with Python*, you will find a quick and easy introduction to Python. If you are new to Python or looking for how to get started with the programming language, reading this appendix will help you get through the initial hurdles.

What you need for this book

You will need the Anaconda Python Distribution to run the examples in this book and write your own Python programs for time series analysis. This is freely downloadable from https://www.continuum.io/downloads.

The code samples of this book have been written using the Jupyter Notebook development environment. To run the Jupyter Notebooks, you need to install Anaconda Python Distribution, which has the Python language essentials, interpreter, packages used to develop the examples, and the Jupyter Notebook server.

Who this book is for

The topics in this book are expected to be useful for the following people:

- Data scientists, professionals with a background in statistics, machine learning, and model building and validation
- Data engineers, professionals with a background in software development
- Software professionals looking to develop an expertise in generating data-driven business insights

Conventions

In this book, you will find a number of text styles that distinguish between different kinds of information. Here are some examples of these styles and an explanation of their meaning.

A block of code is set as follows:

```
import os
import pandas as pd
%matplotlib inline
from matplotlib import pyplot as plt
import seaborn as sns
```

In-text code is highlighted in font and color as here: `pandas.DataFrame`. File and folder names are also shown in the same style, for example, `Chapter_1_Models_for_Time_Series_Analysis.ipynb` or `datasets/DJIA_Jan2016_Dec2016.xlsx`

At several places in the book, we have referred to external URLs to cite source of datasets or other information. A URL would appear in the following text style: `http://finance.yahoo.com`

New terms and **important words** are shown in bold. Words that you see on the screen, for example, in menus or dialog boxes, appear in the text like this: "In order to download new modules, we will go to **Files | Settings | Project Name | Project Interpreter**."

Warnings or important notes appear like this.

Tips and tricks appear like this.

Reader feedback

Feedback from our readers is always welcome. Let us know what you think about this book-what you liked or disliked. Reader feedback is important for us as it helps us develop titles that you will really get the most out of. To send us general feedback, simply email feedback@packtpub.com, and mention the book's title in the subject of your message. If there is a topic that you have expertise in and you are interested in either writing or contributing to a book, see our author guide at www.packtpub.com/authors.

Customer support

Now that you are the proud owner of a Packt book, we have a number of things to help you to get the most from your purchase.

Downloading the example code

You can download the example code files for this book from your account at http://www.packtpub.com. If you purchased this book elsewhere, you can visit http://www.packtpub.com/support and register to have the files emailed directly to you. You can download the code files by following these steps:

1. Log in or register to our website using your email address and password.
2. Hover the mouse pointer on the **SUPPORT** tab at the top.
3. Click on **Code Downloads & Errata**.
4. Enter the name of the book in the **Search** box.
5. Select the book for which you're looking to download the code files.
6. Choose from the drop-down menu where you purchased this book from.
7. Click on **Code Download**.

Once the file is downloaded, please make sure that you unzip or extract the folder using the latest version of:

- WinRAR / 7-Zip for Windows
- Zipeg / iZip / UnRarX for Mac
- 7-Zip / PeaZip for Linux

The code bundle for the book is also hosted on GitHub at `https://github.com/PacktPublishing/Practical-Time-Series-Analysis`. We also have other code bundles from our rich catalog of books and videos available at `https://github.com/PacktPublishing/`. Check them out!

Errata

Although we have taken every care to ensure the accuracy of our content, mistakes do happen. If you find a mistake in one of our books-maybe a mistake in the text or the code- we would be grateful if you could report this to us. By doing so, you can save other readers from frustration and help us improve subsequent versions of this book. If you find any errata, please report them by visiting `http://www.packtpub.com/submit-errata`, selecting your book, clicking on the **Errata Submission Form** link, and entering the details of your errata. Once your errata are verified, your submission will be accepted and the errata will be uploaded to our website or added to any list of existing errata under the Errata section of that title. To view the previously submitted errata, go to `https://www.packtpub.com/books/content/support` and enter the name of the book in the search field. The required information will appear under the **Errata** section.

Piracy

Piracy of copyrighted material on the internet is an ongoing problem across all media. At Packt, we take the protection of our copyright and licenses very seriously. If you come across any illegal copies of our works in any form on the internet, please provide us with the location address or website name immediately so that we can pursue a remedy. Please contact us at `copyright@packtpub.com` with a link to the suspected pirated material. We appreciate your help in protecting our authors and our ability to bring you valuable content.

Questions

If you have a problem with any aspect of this book, you can contact us at `questions@packtpub.com`, and we will do our best to address the problem.

1
Introduction to Time Series

The recent few years have witnessed the widespread application of statistics and machine learning to derive actionable insights and business value out of data in almost all industrial sectors. Hence, it is becoming imperative for business analysts and software professionals to be able to tackle different types of datasets. Often, the data is a time series in the form of a sequence of quantitative observations about a system or process and made at successive points in time. Commonly, the points in time are equally spaced. Examples of time series data include gross domestic product, sales volumes, stock prices, weather attributes when recorded over a time spread of several years, months, days, hours, and so on. The frequency of observation depends on the nature of the variable and its applications. For example, gross domestic product, which is used for measuring annual economic progress of a country, is publicly reported every year. Sales volumes are published monthly, quarterly or biyearly, though figures over longer duration of time might have been generated by aggregating more granular data such as daily or weekly sales. Information about stock prices and weather attributes are available at every second. On the other extreme, there are several physical processes which generate time series data at fraction of a second.

Successful utilization of time series data would lead to monitoring the health of the system over time. For example, the performance of a company is tracked from its quarterly profit margins. Time series analysis aims to utilize such data for several purposes that can be broadly categorized as:

- To understand and interpret the underlying forces that produce the observed state of a system or process over time
- To forecast the future state of the system or process in terms of observable characteristics

To achieve the aforementioned objectives, time series analysis applies different statistical methods to explore and model the internal structures of the time series data such as trends, seasonal fluctuations, cyclical behavior, and irregular changes. Several mathematical techniques and programming tools exist to effectively design computer programs that can explore, visualize, and model patterns in time series data.

However, before taking a deep dive into these techniques, this chapter aims to explain the following two aspects:

- Difference between time series and non-time series data
- Internal structures of time series (some of which have been briefly mentioned in the previous paragraph)

For problem solving, readers would find this chapter useful in order to:

- Distinguish between time series and non-time series data and hence choose the right approach to formulate and solve a given problem.
- Select the appropriate techniques for a time series problem. Depending on the application, one may choose to focus on one or more internal structures of the time series data.

At the end of this chapter, you will understand the different types of datasets you might have to deal with in your analytics project and be able to differentiate time series from non-time series. You will also know about the special internal structures of data which makes it a time series. The overall concepts learnt from this chapter will help in choosing the right approach of dealing with time series.

This chapter will cover the following points:

- Knowing the different types of data you might come across in your analytics projects
- Understanding the internal structures of data that makes a time series
- Dealing with auto-correlation, which is the single most important internal structure of a time series and is often the primary focus of time series analysis

Different types of data

Business analysts and data scientists come across many different types of data in their analytics projects. Most data commonly found in academic and industrial projects can be broadly classified into the following categories:

- Cross-sectional data
- Time series data
- Panel data

Understanding what type of data is needed to solve a problem and what type of data can be obtained from available sources is important for formulating the problem and choosing the right methodology for analysis.

Cross-sectional data

Cross-sectional data or cross-section of a population is obtained by taking observations from multiple individuals at the same point in time. Cross-sectional data can comprise of observations taken at different points in time, however, in such cases time itself does not play any significant role in the analysis. SAT scores of high school students in a particular year is an example of cross-sectional data. Gross domestic product of countries in a given year is another example of cross-sectional data. Data for customer churn analysis is another example of cross-sectional data. Note that, in case of SAT scores of students and GDP of countries, all the observations have been taken in a single year and this makes the two datasets cross-sectional. In essence, the cross-sectional data represents a snapshot at a given instance of time in both the cases. However, customer data for churn analysis can be obtained from over a span of time such as years and months. But for the purpose of analysis, time might not play an important role and therefore though customer churn data might be sourced from multiple points in time, it may still be considered as a cross-sectional dataset.

Often, analysis of cross-sectional data starts with a plot of the variables to visualize their statistical properties such as central tendency, dispersion, skewness, and kurtosis. The following figure illustrates this with the univariate example of military expenditure as a percentage of Gross Domestic Product of 85 countries in the year 2010. By taking the data from a single year we ensure its cross-sectional nature. The figure combines a normalized histogram and a kernel density plot in order to highlight different statistical properties of the military expense data.

As evident from the plot, military expenditure is slightly left skewed with a major peak at roughly around 1.0 %. A couple of minor peaks can also be observed near 6.0 % and 8.0 %.

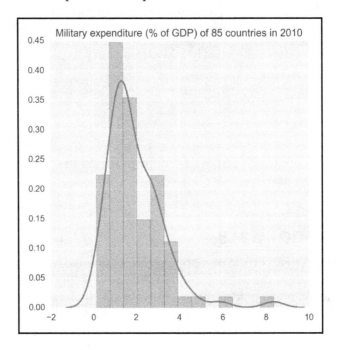

Figure 1.1: Example of univariate cross-sectional data

Exploratory data analysis such as the one in the preceding figure can be done for multiple variables as well in order to understand their joint distribution. Let us illustrate a bivariate analysis by considering total debt of the countries' central governments along with their military expenditure in 2010. The following figure shows the joint distributions of these variables as **kernel density plots**. The bivariate joint distribution shows no clear correlation between the two, except may be for lower values of military expenditure and debt of central government.

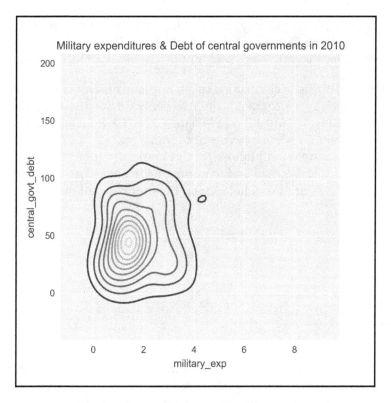

Figure 1.2: Example of bi-variate cross-sectional data

It is noteworthy that analysis of cross-sectional data extends beyond exploratory data analysis and visualization as shown in the preceding example. Advanced methods such as cross-sectional regression fit a linear regression model between several explanatory variables and a dependent variable. For example, in case of customer churn analysis, the objective could be to fit a logistic regression model between customer attributes and customer behavior described by churned or not-churned. The logistic regression model is a special case of generalized linear regression for discrete and binary outcome. It explains the factors that make customers churn and can predict the outcome for a new customer. Since time is not a crucial element in this type of cross-sectional data, predictions can be obtained for a new customer at a future point in time. In this book, we discuss techniques for modeling time series data in which time and the sequential nature of observations are crucial factors for analysis.

The dataset of the example on military expenditures and national debt of countries has been downloaded from the Open Data Catalog of World Bank. You can find the data in the WDIData.csv file under the datasets folder of this book's GitHub repository.

All examples in this book are accompanied by an implementation of the same in Python. So let us now discuss the Python program written to generate the preceding figures. Before we are able to plot the figures, we must read the dataset into Python and familiarize ourselves with the basic structure of the data in terms of columns and rows found in the dataset. Datasets used for the examples and figures, in this book, are in Excel or CSV format. We will use the pandas package to read and manipulate the data. For visualization, matplotlib and seaborn are used. Let us start by importing all the packages to run this example:

```
from __future__ import print_function
import os
import pandas as pd
import numpy as np
%matplotlib inline
from matplotlib import pyplot as plt
import seaborn as sns
```

The print_function has been imported from the __future__ package to enable using print as a function for readers who might be using a 2.x version of Python. In Python 3.x, print is by default a function. As this code is written and executed from an IPython notebook, %matplotlib inline ensures that the graphics packages are imported properly and made to work in the HTML environment of the notebook. The os package is used to set the working directory as follows:

```
os.chdir('D:\Practical Time Series')
```

Now, we read the data from the CSV file and display basic information about it:

```
data = pd.read_csv('datasets/WDIData.csv')
print('Column names:', data.columns)
```

This gives us the following output showing the column names of the dataset:

```
Column names: Index([u'Country Name', u'Country Code', u'Indicator Name',
       u'Indicator Code', u'1960', u'1961', u'1962', u'1963', u'1964',
u'1965',
       u'1966', u'1967', u'1968', u'1969', u'1970', u'1971', u'1972',
u'1973',
       u'1974', u'1975', u'1976', u'1977', u'1978', u'1979', u'1980',
u'1981',
       u'1982', u'1983', u'1984', u'1985', u'1986', u'1987', u'1988',
u'1989',
```

```
        u'1990', u'1991', u'1992', u'1993', u'1994', u'1995', u'1996',
u'1997',
        u'1998', u'1999', u'2000', u'2001', u'2002', u'2003', u'2004',
u'2005',
        u'2006', u'2007', u'2008', u'2009', u'2010', u'2011', u'2012',
u'2013',
        u'2014', u'2015', u'2016'],
      dtype='object')
```

Let us also get a sense of the size of the data in terms of number of rows and columns by running the following line:

```
print('No. of rows, columns:', data.shape)
```

This returns the following output:

```
No. of rows, columns: (397056, 62)
```

This dataset has nearly 400k rows because it captures 1504 world development indicators for 264 different countries. This information about the unique number of indicators and countries can be obtained by running the following four lines:

```
nb_countries = data['Country Code'].unique().shape[0]
print('Unique number of countries:', nb_countries)
```

As it appears from the structure of the data, every row gives the observations about an indicator that is identified by columns Indicator Name and Indicator Code and for the country, which is indicated by the columns Country Name and Country Code. Columns 1960 through 2016 have the values of an indicator during the same period of time. With this understanding of how the data is laid out in the DataFrame, we are now set to extract the rows and columns that are relevant for our visualization.

Let us start by preparing two other DataFrames that get the rows corresponding to the indicators Total Central Government Debt (as % of GDP) and Military expenditure (% of GDP) for all the countries. This is done by slicing the original DataFrame as follows:

```
central_govt_debt = data.ix[data['Indicator Name']=='Central government
debt, total (% of GDP)']
military_exp = data.ix[data['Indicator Name']=='Military expenditure (% of
GDP)']
```

The preceding two lines create two new `DataFrames`, namely `central_govt_debt` and `military_exp`. A quick check about the shapes of these `DataFrames` can be done by running the following two lines:

```
print('Shape of central_govt_debt:', central_govt_debt.shape)
print('Shape of military_exp:', military_exp.shape)
```

These lines return the following output:

```
Shape of central_govt_debt: (264, 62)
Shape of military_exp: (264, 62)
```

These `DataFrames` have all the information we need. In order to plot the univariate and bivariate cross-sectional data in the preceding figure, we need the column `2010`. Before we actually run the code for plotting, let us quickly check if the column `2010` has missing. This is done by the following two lines:

```
central_govt_debt['2010'].describe()
military_exp['2010'].describe()
```

Which generate the following outputs respectively:

```
count       93.000000
mean        52.894412
std         30.866372
min          0.519274
25%               NaN
50%               NaN
75%               NaN
max        168.474953
Name: 2010, dtype: float64
count      194.000000
mean         1.958123
std          1.370594
min          0.000000
25%               NaN
50%               NaN
75%               NaN
max          8.588373
Name: 2010, dtype: float64
```

Which tells us that the describe function could not compute the 25th, 50th, and 75th quartiles for either, hence there are missing values to be avoided.

Additionally, we would like the Country Code column to be the row indices. So the following couple of lines are executed:

```
central_govt_debt.index = central_govt_debt['Country Code']
military_exp.index = military_exp['Country Code']
```

Next, we create two pandas.Series by taking non-empty 2010 columns from central_govt_debt and military_exp. The newly created Series objects are then merged into to form a single DataFrame:

```
central_govt_debt_2010 =
central_govt_debt['2010'].ix[~pd.isnull(central_govt_debt['2010'])]
military_exp_2010 =
military_exp['2010'].ix[~pd.isnull(military_exp['2010'])]
data_to_plot = pd.concat((central_govt_debt_2010, military_exp_2010),
axis=1)
data_to_plot.columns = ['central_govt_debt', 'military_exp']
data_to_plot.head()
```

The preceding lines return the following table that shows that not all countries have information on both Central Government Debt and Military Expense for the year 2010:

	central_govt_debt	military_exp
AFG	NaN	1.897473
AGO	NaN	4.244884
ALB	NaN	1.558592
ARB	NaN	5.122879
ARE	NaN	6.119468
ARG	NaN	0.814878
ARM	NaN	4.265646
ATG	75.289093	NaN
AUS	29.356946	1.951809
AUT	79.408304	0.824770

To plot, we have to take only those countries that have both central government debt and military expense. Run the following line, to filter out rows with missing values:

```
data_to_plot = data_to_plot.ix[(~pd.isnull(data_to_plot.central_govt_debt))
& (~pd.isnull(data_to_plot.military_exp)), :]
```

The first five rows of the filtered `DataFrame` are displayed by running the following line:

```
data_to_plot.head()
```

	central_govt_debt	military_exp
AUS	29.356946	1.951809
AUT	79.408304	0.824770
AZE	6.385576	2.791004
BEL	7.022605	1.084631
BGR	21.286254	1.765384
AUS	29.356946	1.951809
AUT	79.408304	0.824770
AZE	6.385576	2.791004
BEL	7.022605	1.084631
BGR	21.286254	1.765384

The preceding table has only non-empty values and we are now ready to generate the plots for the cross-sectional data. The following lines of code generate the plot on the univariate cross-sectional data on military expense:

```
plt.figure(figsize=(5.5, 5.5))
g = sns.distplot(np.array(data_to_plot.military_exp), norm_hist=False)
g.set_title('Military expenditure (% of GDP) of 85 countries in 2010')
```

The plot is saved as a `png` file under the `plots/ch1` folder of this book's GitHub repository. We will also generate the bivariate plot between military expense and central government debt by running the following code:

```
plt.figure(figsize=(5.5, 5.5))
g = sns.kdeplot(data_to_plot.military_exp,
data2=data_to_plot.central_govt_debt)
g.set_title('Military expenditures & Debt of central governments in 2010')
```

Time series data

The example of cross-sectional data discussed earlier is from the year 2010 only. However, instead if we consider only one country, for example United States, and take a look at its military expenses and central government debt for a span of 10 years from 2001 to 2010, that would get two time series - one about the US federal military expenditure and the other about debt of US federal government. Therefore, in essence, a time series is made up of quantitative observations on one or more measurable characteristics of an individual entity and taken at multiple points in time. In this case, the data represents yearly military expenditure and government debt for the United States. Time series data is typically characterized by several interesting internal structures such as trend, seasonality, stationarity, autocorrelation, and so on. These will be conceptually discussed in the coming sections in this chapter.

The internal structures of time series data require special formulation and techniques for its analysis. These techniques will be covered in the following chapters with case studies and implementation of working code in Python.

The following figure plots the couple of time series we have been talking about:

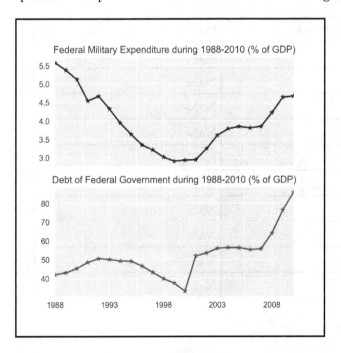

Figure 1.3: Examples of time series data

In order to generate the preceding plots we will extend the code that was developed to get the graphs for the cross-sectional data. We will start by creating two new `Series` to represent the time series of military expenses and central government debt of the United States from 1960 to 2010:

```
central_govt_debt_us = central_govt_debt.ix[central_govt_debt['Country
Code']=='USA', :].T
military_exp_us = military_exp.ix[military_exp['Country Code']=='USA', :].T
```

The two `Series` objects created in the preceding code are merged to form a single `DataFrame` and sliced to hold data for the years 2001 through 2010:

```
data_us = pd.concat((military_exp_us, central_govt_debt_us), axis=1)
index0 = np.where(data_us.index=='1960')[0][0]
index1 = np.where(data_us.index=='2010')[0][0]
data_us = data_us.iloc[index0:index1+1,:]
data_us.columns = ['Federal Military Expenditure', 'Debt of Federal
Government']
data_us.head(10)
```

The data prepared by the preceding code returns the following table:

	Federal Military Expenditure	Debt of Federal Government
1960	NaN	NaN
1961	NaN	NaN
1962	NaN	NaN
1963	NaN	NaN
1964	NaN	NaN
1965	NaN	NaN
1966	NaN	NaN
1967	NaN	NaN
1968	NaN	NaN
1969	NaN	NaN

The preceding table shows that data on federal military expenses and federal debt is not available from several years starting from 1960. Hence, we drop the rows with missing values from the Dataframe data_us before plotting the time series:

```
data_us.dropna(inplace=True)
print('Shape of data_us:', data_us.shape)
```

As seen in the output of the print function, the DataFrame has twenty three rows after dropping the missing values:

```
Shape of data_us: (23, 2)
```

After dropping rows with missing values, we display the first ten rows of data_us are displayed as follows:

	Federal Military Expenditure	Debt of Federal Government
1988	5.57993	42.0258
1989	5.37472	43.1439
1990	5.12025	45.3772
1991	4.53985	48.633
1992	4.66626	50.6016
1993	4.32693	50.1657
1994	3.94129	49.3475
1995	3.63849	49.2366
1996	3.35074	46.7174
1997	3.2099	43.2997

Finally, the time series are generated by executing the following code:

```
# Two subplots, the axes array is 1-d
f, axarr = plt.subplots(2, sharex=True)
f.set_size_inches(5.5, 5.5)
axarr[0].set_title('Federal Military Expenditure during 1988-2010 (% of
GDP)')
data_us['Federal Military Expenditure'].plot(linestyle='-', marker='*',
color='b', ax=axarr[0])
axarr[1].set_title('Debt of Federal Government during 1988-2010 (% of
GDP)')
data_us['Debt of Federal Government'].plot(linestyle='-', marker='*',
color='r', ax=axarr[1])
```

Panel data

So far, we have seen data taken from multiple individuals but at one point in time (cross-sectional) or taken from an individual entity but over multiple points in time (time series). However, if we observe multiple entities over multiple points in time we get a panel data also known as **longitudinal data**. Extending our earlier example about the military expenditure, let us now consider four countries over the same period of 1960-2010. The resulting data will be a panel dataset. The figure given below illustrates the panel data in this scenario. Rows with missing values, corresponding to the period 1960 to 1987 have been dropped before plotting the data.

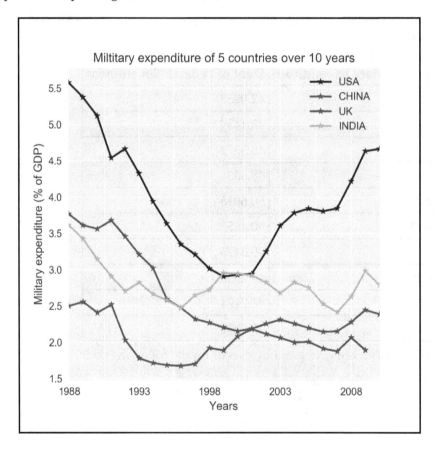

Figure 1.4: Example of panel data

A generic panel data regression model can be stated as $y\_it = W x \_it + b + \epsilon\_it$, which expresses the dependent variable y_it as a linear model of explanatory variable x_it, where W are weights of x_it, b is the bias term, and ϵ_it is the error. i represents individuals for whom data is collected for multiple points in time represented by j. As evident, this type of panel data analysis seeks to model the variations across both multiple individual and multiple points in time. The variations are reflected by ϵ_it and assumptions determine the necessary mathematical treatment. For example, if ϵ_it is assumed to vary non-stochastically with respect to i and t, then it reduces to a dummy variable representing random noise. This type of analysis is known as fixed effects model. On the other hand, ϵ_it varying stochastically over i and t requires a special treatment of the error and is dealt in a random effects model.

Let us prepare the data that is required to plot the preceding figure. We will continue to expand the code we have used for the cross-sectional and time series data previously in this chapter. We start by creating a DataFrame having the data for the four companies mentioned in the preceding plot. This is done as follows:

```
chn = data.ix[(data['Indicator Name']=='Military expenditure (% of GDP)')&\
              (data['Country Code']=='CHN'),index0:index1+1
             ]
chn = pd.Series(data=chn.values[0], index=chn.columns)
chn.dropna(inplace=True)

usa = data.ix[(data['Indicator Name']=='Military expenditure (% of GDP)')&\
              (data['Country Code']=='USA'),index0:index1+1
             ]
usa = pd.Series(data=usa.values[0], index=usa.columns)
usa.dropna(inplace=True)

ind = data.ix[(data['Indicator Name']=='Military expenditure (% of GDP)')&\
              (data['Country Code']=='IND'),index0:index1+1
             ]
ind = pd.Series(data=ind.values[0], index=ind.columns)
ind.dropna(inplace=True)

gbr = data.ix[(data['Indicator Name']=='Military expenditure (% of GDP)')&\
              (data['Country Code']=='GBR'),index0:index1+1
             ]
gbr = pd.Series(data=gbr.values[0], index=gbr.columns)
gbr.dropna(inplace=True)
```

Now that the data is ready for all five countries, we will plot them using the following code:

```
plt.figure(figsize=(5.5, 5.5))
usa.plot(linestyle='-', marker='*', color='b')
chn.plot(linestyle='-', marker='*', color='r')
gbr.plot(linestyle='-', marker='*', color='g')
ind.plot(linestyle='-', marker='*', color='y')
plt.legend(['USA','CHINA','UK','INDIA'], loc=1)
plt.title('Miltitary expenditure of 5 countries over 10 years')
plt.ylabel('Military expenditure (% of GDP)')
plt.xlabel('Years')s
```

 The Jupyter notebook that has the code used for generating all the preceding figures is `Chapter_1_Different_Types_of_Data.ipynb` under the `code` folder in the GitHub repo.

The discussion about different types of data sets the stage for a closer look at time series. We will start doing that by understanding the special properties of data that can be typically found in a time series or panel data with inherent time series in it.

Internal structures of time series

In this section, we will conceptually explain the following special characteristics of time series data that requires its special mathematical treatment:

- General trend
- Seasonality
- Cyclical movements
- Unexpected variations

Most time series has of one or more of the aforementioned internal structures. Based on this notion, a time series can be expressed as $x_t = f_t + s_t + c_t + e_t$, which is a sum of the trend, seasonal, cyclical, and irregular components in that order. Here, t is the time index at which observations about the series have been taken at $t = 1,2,3 ...N$ successive and equally spaced points in time.

The objective of time series analysis is to decompose a time series into its constituent characteristics and develop mathematical models for each. These models are then used to understand what causes the observed behavior of the time series and to predict the series for future points in time.

General trend

When a time series exhibits an upward or downward movement in the long run, it is said to have a general trend. A quick way to check the presence of general trend is to plot the time series as in the following figure, which shows CO_2 concentrations in air measured during 1974 through 1987:

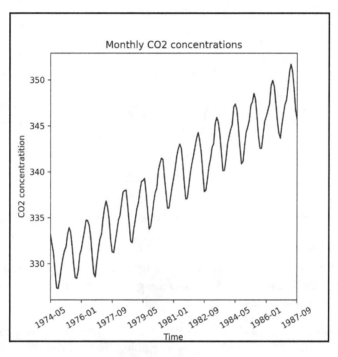

Figure 1.5: Time series of CO2 readings with an upward trend

However, general trend might not be evident over a short run of the series. Short run effects such as seasonal fluctuations and irregular variations cause the time series to revisit lower or higher values observed in the past and hence can temporarily obfuscate any general trend. This is evident in the same time series of CO_2 concentrations when zoomed in over the period of 1979 through 1981, as shown in the following figure. Hence to reveal general trend, we need a time series that dates substantially back in the past.

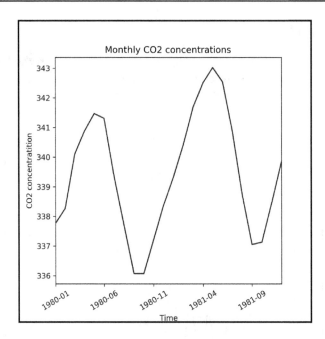

Figure 1.6: Shorter run of CO2 readings time series which is not able to reveal general trend

The general trend in the time series is due to fundamental shifts or systemic changes of the process or system it represents. For example, the upward movement of CO_2 concentrations during 1974 through 1987 can be attributed to the gradual rise in automobiles and industrialization over these years.

A general trend is commonly modeled by setting up the time series as a regression against time and other known factors as explanatory variables. The regression or trend line can then be used as a prediction of the long run movement of the time series. Residuals left by the trend line is further analyzed for other interesting properties such as seasonality, cyclical behavior, and irregular variations.

Now, let us go through the code that generated the preceding plots on CO_2 concentrations. We will also show how to build a trend model using linear regression on the time index (which in this case is the index of the year in the data) as explanatory variable and the CO_2 concentration as the dependent variable. But first, let us load the data in a `pandas.DataFrame`.

The data for this example is in the Excel file
Monthly_CO2_Concentrations.xlsx under the datasets folder of the
GitHub repo.

We start by importing the required packages as follows:

```
from __future__ import print_function
import os
import pandas as pd
import numpy as np
%matplotlib inline
from matplotlib import pyplot as plt
import seaborn as sns
os.chdir('D:\Practical Time Series')
data = pd.read_excel('datasets/Monthly_CO2_Concentrations.xlsx',
converters={'Year': np.int32, 'Month':  np.int32})
data.head()
```

We have passed the argument converters to the read_excel function in order to make
sure that columns Year and Month are assigned the integer (np.int32) datatype. The
preceding lines of code will generate the following table:

	CO$_2$	Year	Month
0	333.13	1974	5
1	332.09	1974	6
2	331.10	1974	7
3	329.14	1974	8
4	327.36	1974	9

Before plotting we must remove all columns having missing values. Besides, the
DataFrame is sorted in ascending order of Year and Month. These are done as follows:

```
data = data.ix[(~pd.isnull(data['CO2']))&\
               (~pd.isnull(data['Year']))&\
               (~pd.isnull(data['Month'])))]
data.sort_values(['Year', 'Month'], inplace=True)
```

Finally, the plot for the time period 1974 to 1987 is generated by executing the following
lines:

```
plt.figure(figsize=(5.5, 5.5))
data['CO2'].plot(color='b')
```

```
plt.title('Monthly CO2 concentrations')
plt.xlabel('Time')
plt.ylabel('CO2 concentratition')
plt.xticks(rotation=30)
```

The zoomed-in version of the data for the time period 1980 to 1981 is generated by after the DataFrame for these three years:

```
plt.figure(figsize=(5.5, 5.5))
data['CO2'].loc[(data['Year']==1980) |
(data['Year']==1981)].plot(color='b')
plt.title('Monthly CO2 concentrations')
plt.xlabel('Time')
plt.ylabel('CO2 concentratition')
plt.xticks(rotation=30)
```

Next, let us fit the trend line. For this we import the LinearRegression class from scikit-learn and fit a linear model on the time index:

```
from sklearn.linear_model import LinearRegression
trend_model = LinearRegression(normalize=True, fit_intercept=True)
trend_model.fit(np.array(data.index).reshape((-1,1)), data['CO2'])
print('Trend model coefficient={} and
intercept={}'.format(trend_model.coef_[0],
trend_model.intercept_)
     )
```

This produces the following output:

```
Trend model coefficient=0.111822078545 and intercept=329.455422234
```

The residuals obtained from the trend line model are shown in the following figure and appear to have seasonal behaviour, which is discussed in the next sub section.

The residuals are calculated and plotted in the preceding by the following lines of code:

```
residuals = np.array(data['CO2']) -
trend_model.predict(np.array(data.index).reshape((-1,1)))
plt.figure(figsize=(5.5, 5.5))
pd.Series(data=residuals, index=data.index).plot(color='b')
plt.title('Residuals of trend model for CO2 concentrations')
plt.xlabel('Time')
plt.ylabel('CO2 concentratition')
plt.xticks(rotation=30)
```

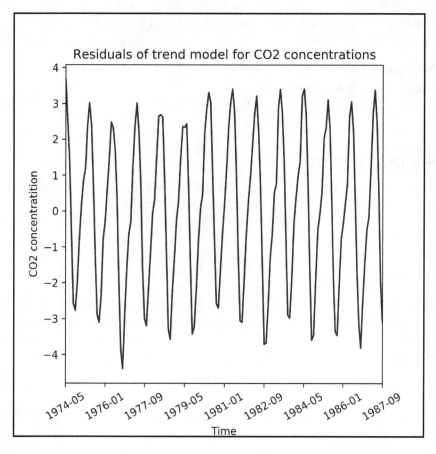

Figure 1.7: Residuals from a linear model of the general trend in CO2 readings

Seasonality

Seasonality manifests as repetitive and period variations in a time series. In most cases, exploratory data analysis reveals the presence of seasonality. Let us revisit the de-trended time series of the CO_2 concentrations. Though the de-trended line series has constant mean and constant variance, it systematically departs from the trend model in a predictable fashion.

Seasonality is manifested as periodic deviations such as those seen in the de-trended observations of CO_2 emissions. Peaks and troughs in the monthly sales volume of seasonal goods such as Christmas gifts or seasonal clothing is another example of a time series with seasonality.

A practical technique of determining seasonality is through exploratory data analysis through the following plots:

- Run sequence plot
- Seasonal sub series plot
- Multiple box plots

Run sequence plot

A simple run sequence plot of the original time series with time on x-axis and the variable on y-axis is good for indicating the following properties of the time series:

- Movements in mean of the series
- Shifts in variance
- Presence of outliers

The following figure is the run sequence plot of a hypothetical time series that is obtained from the mathematical formulation $x_t = (At + B) sin(t) + \epsilon(t)$ with a time-dependent mean and error $\epsilon(t)$ that varies with a normal distribution $N(0, at + b)$ variance. Additionally, a few exceptionally high and low observations are also included as outliers.

In cases such as this, a run sequence plot is an effective way of identifying shifting mean and variance of the series as well as outliers. The plot of the de-trended time series of the CO2 concentrations is an example of a run sequence plot.

Seasonal sub series plot

For a known periodicity of seasonal variations, seasonal sub series redraws the original series over batches of successive time periods. For example, the periodicity in the CO_2 concentrations is 12 months and based on this a seasonal sub series plots on mean and standard deviation of the residuals are shown in the following figure. To visualize seasonality in the residuals, we create quarterly mean and standard deviations.

A seasonal sub series reveals two properties:

- Variations within seasons as within a batch of successive months
- Variations between seasons as between batches of successive months

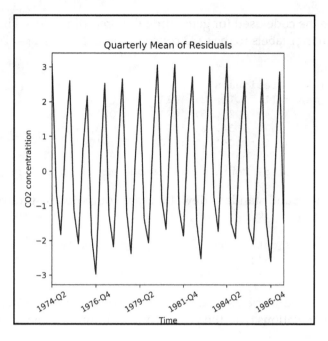

Figure 1.8: Quarterly mean of the residuals from a linear model of the general trend in CO2 readings

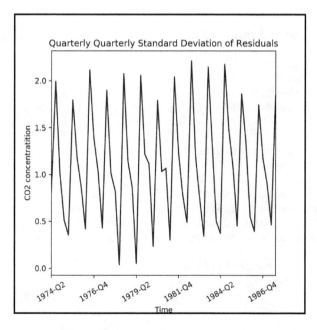

Figure 1.9: Quarterly standard deviation of the residuals from a linear model of the general trend in CO2 readings

Let us now describe the code used for generating the preceding plots. First, we need to add the residuals and quarter labels to the CO2 concentrations `DataFrame`. This is done as follows:

```
data['Residuals'] = residuals
month_quarter_map = {1: 'Q1', 2: 'Q1', 3: 'Q1',
                     4: 'Q2', 5: 'Q2', 6: 'Q2',
                     7: 'Q3', 8: 'Q3', 9: 'Q3',
                     10: 'Q4', 11: 'Q4', 12: 'Q4'
                     }
data['Quarter'] = data['Month'].map(lambda m: month_quarter_map.get(m))
```

Next, the seasonal mean and standard deviations are computed by grouping by the data over `Year` and `Quarter`:

```
seasonal_sub_series_data = data.groupby(by=['Year',
'Quarter'])['Residuals']\
                        .aggregate([np.mean, np.std])
```

This creates the new `DataFrame` as `seasonal_sub_series_data`, which has quarterly mean and standard deviations over the years. These columns are renamed as follows:

```
seasonal_sub_series_data.columns = ['Quarterly Mean', 'Quarterly Standard
Deviation']
```

Next, the quarterly mean and standard deviations are plotted by running the following lines of code:

```
#plot quarterly mean of residuals
plt.figure(figsize=(5.5, 5.5))
seasonal_sub_series_data['Quarterly Mean'].plot(color='b')
plt.title('Quarterly Mean of Residuals')
plt.xlabel('Time')
plt.ylabel('CO2 concentratition')
plt.xticks(rotation=30)

#plot quarterly standard deviation of residuals
plt.figure(figsize=(5.5, 5.5))
seasonal_sub_series_data['Quarterly Standard Deviation'].plot(color='b')
plt.title('Quarterly Standard Deviation of Residuals')
plt.xlabel('Time')
plt.ylabel('CO2 concentratition')
plt.xticks(rotation=30)
```

Multiple box plots

The seasonal sub series plot can be more informative when redrawn with seasonal box plots as shown in the following figure. A box plot displays both central tendency and dispersion within the seasonal data over a batch of time units. Besides, separation between two adjacent box plots reveal the within season variations:

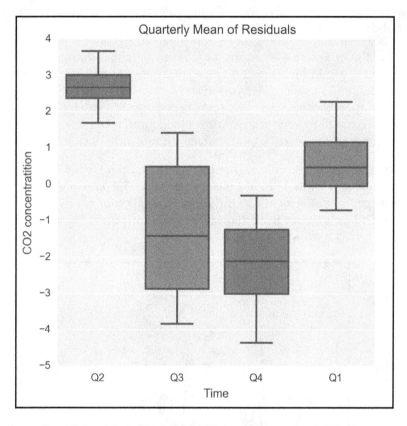

Figure 1.10: Quarterly boxplots of the residuals from a linear model of the general trend in CO2 readings

The code for generating the box plots is as follows:

```
plt.figure(figsize=(5.5, 5.5))
g = sns.boxplot(data=data, y='Residuals', x='Quarter')
g.set_title('Quarterly Mean of Residuals')
g.set_xlabel('Time')
g.set_ylabel('CO2 concentratition')
```

Cyclical changes

Cyclical changes are movements observed after every few units of time, but they occur less frequently than seasonal fluctuations. Unlike seasonality, cyclical changes might not have a fixed period of variations. Besides, the average periodicity for cyclical changes would be larger (most commonly in years), whereas seasonal variations are observed within the same year and corresponds to annual divisions of time such as seasons, quarters, and periods of festivity and holidays and so on.

A long run plot of the time series is required to identify cyclical changes that can occur, for example, every few years and manifests as repetitive crests and troughs. In this regard, time series related to economics and business often show cyclical changes that correspond to usual business and macroeconomic cycles such as periods of recessions followed by every of boom, but are separated by few years of time span. Similar to general trends, identifying cyclical movements might require data that dates significantly back in the past.

The following figure illustrates cyclical changes occurring in inflation of **consumer price index (CPI)** for India and United States over the period of 1960 through 2016. Both the countries exhibit cyclical patterns in CPI inflation, which is roughly over a period of 2-2.5 years. Moreover, CPI inflation of India has larger variations pre-1990 than that seen after 1990.

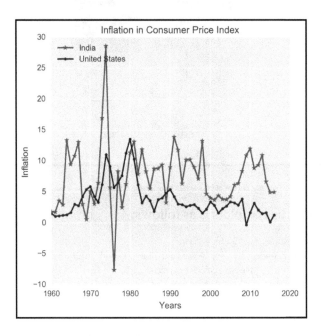

Figure 1.11: Example of cyclical movements in time series data

Source: The data for the preceding figure has been downloaded from `http://datamarket.` `com`, which maintains data on time series from a wide variety of subjects.

 You can find the CPI inflation dataset in file `inflation-consumer-` `prices-annual.xlsx`, which is in the `datasets`folder, on the GitHub repository.

The code written to generate the figure is as follows:

```
inflation = pd.read_excel('datasets/inflation-consumer-prices-annual.xlsx',
parse_dates=['Year'])
plt.figure(figsize=(5.5, 5.5))
plt.plot(range(1960,2017), inflation['India'], linestyle='-', marker='*',
color='r')
plt.plot(range(1960,2017), inflation['United States'], linestyle='-',
marker='.', color='b')
plt.legend(['India','United States'], loc=2)
plt.title('Inflation in Consumer Price Index')
plt.ylabel('Inflation')
plt.xlabel('Years')
```

Unexpected variations

Referring to our model that expresses a time series as a sum of four components, it is noteworthy that in spite of being able to account for the three other components, we might still be left with an irreducible error component that is random and does not exhibit systematic dependency on the time index. This fourth component reflects unexpected variations in the time series. Unexpected variations are stochastic and cannot be framed in a mathematical model for a definitive future prediction. This type of error is due to lack of information about explanatory variables that can model these variations or due to presence of a random noise.

 The IPython notebook for all the code developed in this section is in the file `Chapter_1_Internal_Structures.ipynb` in the `code` folder of this book's GitHub repository.

Models for time series analysis

The purpose of time series analysis is to develop a mathematical model that can explain the observed behavior of a time series and possibly forecast the future state of the series. The chosen model should be able to account for one or more of the internal structures that might be present. To this end, we will give an overview of the following general models that are often used as building blocks of time series analysis:

- Zero mean models
- Random walk
- Trend models
- Seasonality models

Zero mean models

The zero-mean models have a constant mean and constant variance and shows no predictable trends or seasonality. Observations from a zero mean model are assumed to be **independent and identically distributed** (**iid**) and represent the random noise around a fixed mean, which has been deducted from the time series as a constant term.

Let us consider that X_1, X_2, ... ,X_n represent the random variables corresponding to n observations of a zero mean model. If x_1, x_2, ... ,x_n are n observations from the zero mean time series, then the joint distribution of the observations is given as a product of probability mass function for every time index as follows:

$$P(X1 = x1, X2 = x2, ... , Xn = xn) = f(X1 = x1)\, f(X2 = x2) ... f(Xn = xn)$$

Most commonly $f(X_t = x_t)$ is modeled by a normal distribution of mean zero and variance σ^2, which is assumed to be the irreducible error of the model and hence treated as a random noise. The following figure shows a zero-mean series of normally distributed random noise of unit variance:

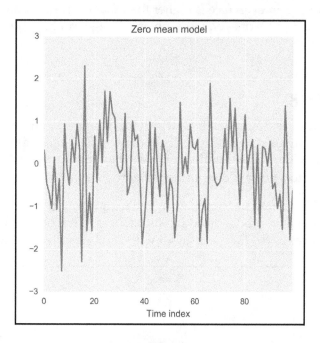

Figure 1.12: Zero-mean time series

The preceding plot is generated by the following code:

```
import os
import numpy as np
%matplotlib inline
from matplotlib import pyplot as plt
import seaborn as sns
os.chdir('D:/Practical Time Series/')
zero_mean_series = np.random.normal(loc=0.0, scale=1., size=100)
```

The zero mean with constant variance represents a random noise that can assume infinitely possible real values and is suited for representing irregular variations in the time series of a continuous variable. However in many cases, the observable state of the system or process might be discrete in nature and confined to a finite number of possible values $s_1, s_2, ..., s_m$. In such cases, the observed variable (X) is assumed to obey the multinomial distribution, $P(X = s_1) = p_1$, $P(X = s_2) = p_2,...,P(X = s_m) = p_m$ such that $p_1 + p_2 + ... + p_m = 1$. Such a time series is a discrete stochastic process.

Multiple throws a dice over time is an example of a discrete stochastic process with six possible outcomes for any throw. A simpler discrete stochastic process is a binary process such as tossing a coin such as only two outcomes namely head and tail. The following figure shows 100 runs from a simulated process of throwing a biased dice for which probability of turning up an even face is higher than that of showing an odd face. Note the higher number of occurrences of even faces, on an average, compared to the number of occurrences of odd faces.

Random walk

A random walk is given as a sum of n iids, which has zero mean and constant variance. Based on this definition, the realization of a random walk at time index t is given by the sum $S = x_1 + x_2 + ... + x_n$. The following figure shows the random walk obtained from iids, which vary according to a normal distribution of zero mean and unit variance.

The random walk is important because if such behavior is found in a time series, it can be easily reduced to zero mean model by taking differences of the observations from two consecutive time indices as $S_t - S_{t-1} = x_t$ is an iid with zero mean and constant variance.

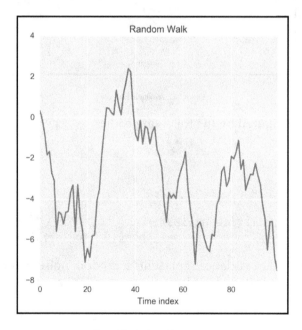

Figure 1.13: Random walk time series

The random walk in the preceding figure can be generated by taking the cumulative sum of the zero mean model discussed in the previous section. The following code implements this:

```
random_walk = np.cumsum(zero_mean_series)
plt.figure(figsize=(5.5, 5.5))
g = sns.tsplot(random_walk)
g.set_title('Random Walk')
g.set_xlabel('Time index')
```

Trend models

This type of model aims to capture the long run trend in the time series that can be fitted as linear regression of the time index. When the time series does not exhibit any periodic or seasonal fluctuations, it can be expressed just as the sum of the trend and the zero mean model as $x_t = \mu(t) + y_t$, where $\mu(t)$ is the time-dependent long run trend of the series.

The choice of the trend model $\mu(t)$ is critical to correctly capturing the behavior of the time series. Exploratory data analysis often provides hints for hypothesizing whether the model should be linear or non-linear in t. A linear model is simply $\mu(t) = wt + b$, whereas quadratic model is $\mu(t) = w_1t + w_2t^2 + b$. Sometimes, the trend can be hypothesized by a more complex relationship in terms of the time index such as $\mu(t) = w_0t^p + b$.

The weights and biases in the trend modes such as the ones discussed previously is obtained by running a regression with t as the explanatory variable and μ as the explained. The residuals $x_t - \mu(t)$ of the trend model is considered to the irreducible noise and as realization of the zero mean component y_t.

Seasonality models

Seasonality manifests as periodic and repetitive fluctuations in a time series and hence are modeled as sum of weighted sum of sine waves of known periodicity. Assuming that long run trend has been removed by a trend line, the seasonality model can be expressed as $x_t = s_t + y_t$, where the seasonal variation $s_t \sum \frac{L}{k=1} (w_k \cos\alpha t + v_k \sin\alpha t) + b$ with known periodicity is α.

Seasonality models are also known as **harmonic regression model** as they attempt to fit the sum of multiple sin waves.

The four models described here are building blocks of a fully-fledged time series model. As you might have gathered by now, a zero sum model represents irreducible error of the system and all of other three models aim to transform a given time series to the zero sum models through suitable mathematical transformations. To get forecasts in terms of the original time series, relevant inverse transformations are applied.

The upcoming chapters detail the four models discussed here. However, we have reached a point where we can summarize the generic approach of a time series analysis in the following four steps:

- Visualize the data at different granularities of the time index to reveal long run trends and seasonal fluctuations
- Fit trend line capture long run trends and plot the residuals to check for seasonality or irreducible error
- Fit a harmonic regression model to capture seasonality
- Plot the residuals left by the seasonality model to check for irreducible error

These steps are most commonly enough to develop mathematical models for most time series. The individual trend and seasonality models can be simple or complex depending on the original time series and the application.

 The code written in this section can be found in the `Chapter_1_Models_for_Time_Series_Analysis.ipynb` IPython notebook located in the `code` folder of this book's GitHub repository.

Autocorrelation and Partial autocorrelation

After applying the mathematical transformations discussed in the previous section, we will often be left with what is known as a **stationary** (or weakly stationary) **time series**, which is characterized by a constant mean $E(x_t)$ and correlation that depends only on the time lag between two time steps, but independent of the value of the time step. This type of covariance is the key in time series analysis and is called **autocovariance** or **autocorrelation** when normalized to the range of -1 to 1. Autocorrelation is therefore expressed as the second order moment $E(x_t, x_{t+h}) = g(h)$ that evidently is a function of only the time lag h and independent of the actual time index t. This special definition of autocorrelation ensures that it is a time-independent property and hence can be reliably used for making inference about future realization of the time series.

Autocorrelation reflects the degree of linear dependency between the time series at index t and the time series at indices t-h or t+h. A positive autocorrelation indicates that the present and future values of the time series move in the same direction, whereas negative values means that present and future values move in the opposite direction. If autocorrelation is close to zero, temporal dependencies within the series may be hard to find. Because of this property, autocorrelation is useful in predicting the future state of a time series at h time steps ahead.

Presence of autocorrelation can be identified by plotting the observed values of the **autocorrelation function** (**ACF**) for a given time series. This plot is commonly referred as the ACF plot. Let us illustrate how plotting the observed values of the ACF can help in detecting presence of autocorrelation. For this we first plot the daily value of **Dow Jones Industrial Average** (**DJIA**) observed during January 2016 to December 2016:

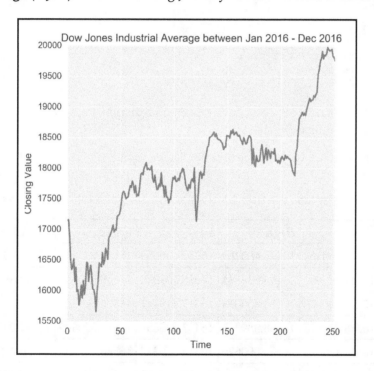

Figure 1.14: Time series of Dow Jones Industrial Average

From the preceding figure, it might be apparent that when DJIA starts rising, it continues to rise for some time and vice-versa. However, we must ascertain this through an ACF plot.

 The dataset for this plot has been downloaded from Yahoo! Finance (`http://finance.yahoo.com`) and kept as `DJIA_Jan2016_Dec2016.xlsx` under the `datasets` folder of this book's GitHub repository.

We will use `pandas` to read data from the Excel file and seaborn along with matplotlib to visualize the time series. Like before, we will also use the os package to set the working directory.

So let us first import these packages:

```
import os
import pandas as pd
%matplotlib inline
from matplotlib import pyplot as plt
import seaborn as sns
os.chdir('D:/Practical Time Series')
```

Next, we load the data as a `pandas.DataFrame` and display its first 10 rows to have a look at the columns of the dataset:

```
djia_df = pd.read_excel('datasets/DJIA_Jan2016_Dec2016.xlsx')
djia_df.head(10)
```

The preceding code displays the first 10 rows of the dataset as shown in the following table:

	Date	Open	High	Low	Close	Adj Close	Volume
0	2016-01-04	17405.480469	17405.480469	16957.630859	17148.939453	17148.939453	148060000
1	2016-01-05	17147.500000	17195.839844	17038.609375	17158.660156	17158.660156	105750000
2	2016-01-06	17154.830078	17154.830078	16817.619141	16906.509766	16906.509766	120250000
3	2016-01-07	16888.359375	16888.359375	16463.630859	16514.099609	16514.099609	176240000
4	2016-01-08	16519.169922	16651.890625	16314.570313	16346.450195	16346.450195	141850000
5	2016-01-11	16358.709961	16461.849609	16232.030273	16398.570313	16398.570313	127790000
6	2016-01-12	16419.109375	16591.349609	16322.070313	16516.220703	16516.220703	117480000
7	2016-01-13	16526.630859	16593.509766	16123.200195	16151.410156	16151.410156	153530000
8	2016-01-14	16159.009766	16482.050781	16075.120117	16379.049805	16379.049805	158830000
9	2016-01-15	16354.330078	16354.330078	15842.110352	15988.080078	15988.080078	239210000

The first column in the preceding table is always the default row index created by the `pandas.read_csv` function.

We have used the closing value of DJIA, which is given in column `Close`, to illustrate autocorrelation and the ACF function. The time series plot has been generated as follows:

```
plt.figure(figsize=(5.5, 5.5))
g = sns.tsplot(djia_df['Close'])
g.set_title('Dow Jones Industrial Average between Jan 2016 - Dec 2016')
g.set_xlabel('Time')
g.set_ylabel('Closing Value')
```

Next, the ACF is estimated by computing autocorrelation for different values of lag h, which in this case is varied from 0 through 30. The `Pandas.Series.autocorr` function is used to calculate the autocorrelation for different values the lag. The code for this is given as follows:

```
lag = range(0,31)
    djia_acf = []
for l in lag:
    djia_acf.append(djia_df['Close'].autocorr(l))
```

The preceding code, iterates over a list of 31 values of the lag starting from 0 to 30. A lag of 0 indicates autocorrelation of an observation with itself (in other words self-correlation) and hence it is expected to be 1.0 as also confirmed in the following figure. Autocorrelation in DJIA Close values appears to linearly drop with the lag with an apparent change in the rate of the drop at around 18 days. At a lag of 30 days the ACF is a bit over 0.65.

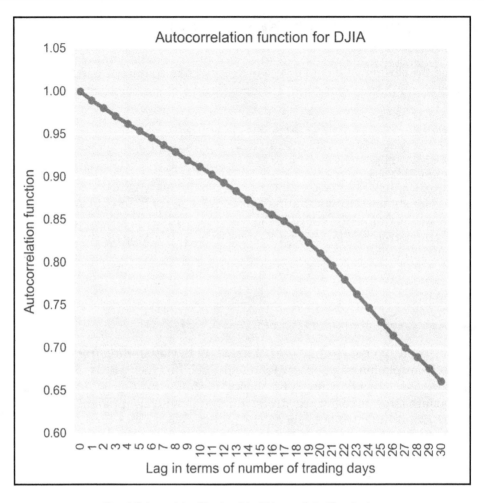

Figure 1.15: Autocorrelation of Dow Jones Industrial Average calculated for various lags

The preceding plot has been generated by the following code:

```
plt.figure(figsize=(5.5, 5.5))
g = sns.pointplot(x=lag, y=djia_acf, markers='.')
g.set_title('Autocorrelation function for DJIA')
g.set_xlabel('Lag in terms of number of trading days')
g.set_ylabel('Autocorrelation function')
g.set_xticklabels(lag, rotation=90)
plt.savefig('plots/ch2/B07887_02_11.png', format='png', dpi=300)
```

The ACF plot shows that autocorrelation, in the case of DJIA Close values, has a functional dependency on the time lag between observations.

> The code developed to run the analysis in this section is in the IPyton notebook `Chapter1_Autocorrelation.ipynb` under the `code` folder of the GitHub repository of this book.

We have written a for-loop to calculate the autocorrelation at different lags and plotted the results using the `sns.pointplot` function. Alternatively, the `plot_acf` function of `statsmodels.graphics.tsaplots` to compute and plot the autocorrelation at various lags. Additionally, this function also plots the 95% confidence intervals. Autocorrelation outside these confidence intervals is statistically significant correlation while those which are inside the confidence intervals are due to random noise.

The autocorrelation and confidence intervals generated by the `plot_acf` is shown in the following figure:

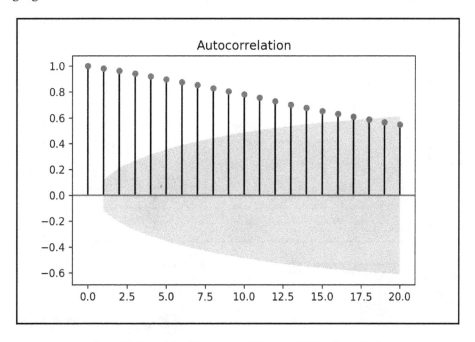

Figure 1.16: Autocorrelation of Dow Jones Industrial Average with 95% confidence intervals

So far, we have discussed autocorrelation which is a measure of linear dependency between variables x_t and $x_{(t+h)}$. **Autoregressive (AR)** models captures this dependency as a linear regression between the $x_{(t+h)}$ and x_t. However, time series tend to carry information and dependency structures in steps and therefore autocorrelation at lag h is also influenced by the intermediate variables $x_t, x_{(t+1)}...x_{(t+h-1)}$. Therefore, autocorrelation is not the correct measure of the mutual correlation between x_t and $x_{(t+h)}$ in the presence of the intermediate variables. Hence, it would erroneous to choose h in AR models based on autocorrelation. Partial autocorrelation solves this problem by measuring the correlation between x_t and $x_{(t+h)}$ when the influence of the intermediate variables has been removed. Hence partial autocorrelation in time series analysis defines the correlation between x_t and $x_{(t+h)}$ which is not accounted for by lags $t+1$ to $t+h-1$.

Partial autocorrelation helps in identifying the order h of an AR(h) model. Let us plot the partial autocorrelation of DJIA Close Values using *plot_pacf* as follows:

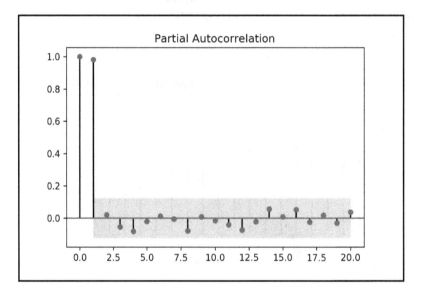

Figure 1.17: Partial autocorrelation of Dow Jones Industrial Average with 95% confidence intervals

The first partial autocorrelation at lag zero is always 1.0. As seen in the preceding figure, the partial autocorrelation only at lag one is statistically significant while for rest the lags it is within the 95% confidence intervals. Hence, for DJIA Close Values the order of AR models is one.

Summary

In this chapter, we discussed several types of data such as cross-sectional, time series, and panel data. We delved into the special properties that make time series data special. Several examples and working code in Python have been discussed to give an understanding of how exploratory data analysis can be performed on time series to visualize its properties. We also described autocorrelation and partial autocorrelation and graphical techniques to detect these in a time series. The topics discussed in this chapter give us the stage for a more detailed discussion for working on time series data in Python.

In the next chapter, you will learn how to read more complex data types in time series and use such information for more in-depth exploratory data analysis. We will revisit autocorrelation in the context of stationarity of time series. Statistical methods to detect autocorrelation would be discussed. We would also discuss importance of stationarity and describe different differencing and averaging methods for stationarizing a non-stationary time series. Additive and multiplicative models of time decomposition for estimating trend and seasonality are discussed.

2
Understanding Time Series Data

In the previous chapter, we touched upon a general approach of time series analysis which consists of two main steps:

- Data visualization to check the presence of trend, seasonality, and cyclical patterns
- Adjustment of trend and seasonality to generate stationary series

Generating stationary data is important for enhancing the time series forecasting model. Deduction of the trend, seasonal, and cyclical components would leave us with irregular fluctuations which cannot be modeled by using only the time index as an explanatory variable. Therefore, in order to further improve forecasting, the irregular fluctuations are assumed to be **independent and identically distributed** (**iid**) observations and modeled by a linear regression on variables other than the time index.

For example, house prices might exhibit both trend and seasonal (for example, quarterly) variations. However, the residuals left after adjusting trend and seasonality might actually depend on exogenous variables, such as total floor space, number of floors in the building and so on, which depend on specific instances in the sampled data. Therefore trend and seasonality adjustments along with a model on exogenous variables would be a better forecast for future instances of the time series.

Changing the original time series to iid observations, or in other words stationarizing a time series, is an important step for developing the linear regression model on exogenous variables. This is because there exist well-established statistical methods, for example central limit theorem, least squares method, and so on, which work well for iid observations.

The methodology for time series analysis described in the preceding paragraphs is summarized in the following flowchart. In this chapter, we will cover steps 1, 2, and 3 of this methodology in the following topics:

- Advanced processing and visualization of time series data
- Statistical hypothesis testing to verify stationarity of a time series
- Time series decomposition for adjusting trends and seasonality

The reader would find the mathematical concepts discussed in this chapter to be fundamental building blocks for developing time series forecasting models.

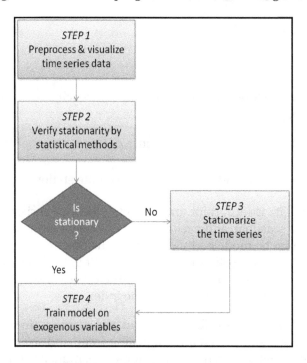

Figure 2.1: A generic methodology of time series analysis

Advanced processing and visualization of time series data

In many cases, the original time series needs to be transformed into aggregate statistics. For example, observations in the original time series might have been recorded at every second; however, in order to perform any meaningful analysis, data must be aggregated every minute. This would need resampling the observations over periods that are longer than the granular time indices in the original data. The aggregate statistics, such as mean, median, and variance, is calculated for each of the longer periods of time.

Another example of data pre-processing for time series, is computing aggregates over similar segments in the data. Consider the monthly sales of cars manufactured by company X where the data exhibits monthly seasonality, due to which sales during a month of a given year shows patters similar to the sales of the same month in the previous and next years. To highlight this kind of seasonality we must remove the long-run trend from the data. However, let us assume there is no long-run trend or it has already been adjusted. We are interested in estimating the season wise (in this case month wise) statistics to determine between-seasons (between-months) variations.

For example, we require average sales during January, February, and so on to understand how the sales vary on an average over any given year. Season wise trends, such as month wise sales of cars, can be obtained by first grouping the original time series into 12 segments, each segment being defined by a month, and then aggregating the data for every segment.

Note that both resampling and group-by operations divide the original time series in non-overlapping chunks to find the aggregates. Both the techniques are going to reduce noise and subsequently produce smoothing of the original time series.

However, sometimes continuous or running aggregates of the time series are required for analysis. The technique of computing aggregates over windows of successive time periods gives moving or rolling aggregates. For example, quarterly moving averages of the car sales data would be finding averages over a four month window that shifts by one month every time. By shifting or rolling the window of computation, the moving averages are generated.

The techniques are demonstrated with examples in the following three sub sections:

- Resampling time series data
- Performing `group-by`
- Calculating moving statistics

We will use the pandas data processing API for implementing the techniques.

Resampling time series data

The technique of resmapling is illustrated using a time series on chemical concentration readings taken every two hours between 1ˢᵗ January 1975 and 17ᵗʰ January 1975. The dataset has been downloaded from `http://datamarket.com` and is also available in the datasets folder of this book's GitHub repo.

We start by importing the packages required for running this example:

```
from __future__ import print_function
import os
import pandas as pd
import numpy as np
%matplotlib inline
from matplotlib import pyplot as plt
```

Then we set the working directory as follows:

```
os.chdir('D:/Practical Time Series')
```

This is followed by reading the data from the CSV file in a `pandas.DataFrame` and displaying shape and the first 10 rows of the `DataFrame`:

```
df = pd.read_csv('datasets/chemical-concentration-readings.csv')
print('Shape of the dataset:', df.shape)
df.head(10)
```

The preceding code returns the following output:

```
Shape of the dataset: (197, 2)
```

	Timestamp	Chemical conc.
0	1975-01-01 00:00:00	17.0
1	1975-01-01 02:00:00	16.6
2	1975-01-01 04:00:00	16.3
3	1975-01-01 06:00:00	16.1
4	1975-01-01 08:00:00	17.1
5	1975-01-01 10:00:00	16.9

6	1975-01-01 12:00:00	16.8
7	1975-01-01 14:00:00	17.4
8	1975-01-01 16:00:00	17.1
9	1975-01-01 18:00:00	17.0

We will convert the bi-hourly observations of the original time series to daily averages by applying the resample and mean functions on the second column. The resample function requires the row indices of the DataFrame to be timestamp of type numpy.datetime64. Therefore, we change the row indices from whole numbers, as shown in the preceding table, to datetime_rowid which is a pandas.Series of numpy.datetime64 objects. The numpy.datetime64 objects are generated from the Timestamp column by using the pd.to datetime utility function. The following code shows how the row wise re-indexing is done:

```
datetime_rowid = df['Timestamp'].map(lambda t: pd.to_datetime(t,
format='%Y-%m-%d %H:%M:%S'))
df.index = datetime_rowid
df.head(10)
```

The last line returns the first ten rows of the re-indexed DataFrame as follows. Notice that the new row indices are of type numpy.datetime64.

Timestamp	Timestamp	Chemical conc.
1975-01-01 00:00:00	1975-01-01 00:00:00	17.0
1975-01-01 02:00:00	1975-01-01 02:00:00	16.6
1975-01-01 04:00:00	1975-01-01 04:00:00	16.3
1975-01-01 06:00:00	1975-01-01 06:00:00	16.1
1975-01-01 08:00:00	1975-01-01 08:00:00	17.1
1975-01-01 10:00:00	1975-01-01 10:00:00	16.9
1975-01-01 12:00:00	1975-01-01 12:00:00	16.8
1975-01-01 14:00:00	1975-01-01 14:00:00	17.4
1975-01-01 16:00:00	1975-01-01 16:00:00	17.1
1975-01-01 18:00:00	1975-01-01 18:00:00	17.0

Now we are ready to apply the `resample` and `mean` functions on the `Chemical conc.` column:

```
daily = df['Chemical conc.'].resample('D')
daily_mean = daily.mean()
```

Notice that we have passed the argument `D` to the `resample` function to generate the daily averages. For monthly and yearly aggregates, we need to pass `M` and `Y` to the `resample` function.

Finally, the original and the daily averages are plotted in the following figure, which shows the smoothing effect of the latter:

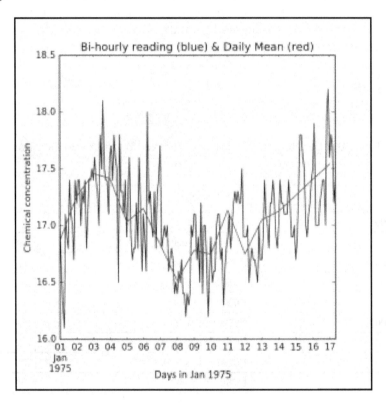

Figure 2.2: Time series of bi-hourly readings and daily mean of chemical concentrations

The code that generates the preceding figure and saves it as a PNG file is given as follows:

```
fig = plt.figure(figsize=(5.5, 5.5))
ax = fig.add_subplot(1,1,1)
```

```
df['Chemical conc.'].plot(ax=ax, color='b')
daily_mean.plot(ax=ax, color='r')

ax.set_title('Bi-hourly reading (blue) & Daily Mean (red)')
ax.set_xlabel('Day in Jan 1975')
ax.set_ylabel('Chemical concentration')
plt.savefig('plots/ch2/B07887_02_17.png', format='png', dpi=300)
```

Group wise aggregation

To demonstrate group wise aggregation we will use the time series on mean daily temperature of Fisher River in Texas, United States. The time series has observations taken between 1st January 1988 and 31st December 1991. The dataset has been downloaded from `http://datamarket.com` and is also available in the datasets folder of this book's GitHub repo.

We start by reading and plotting the original time series as follows:

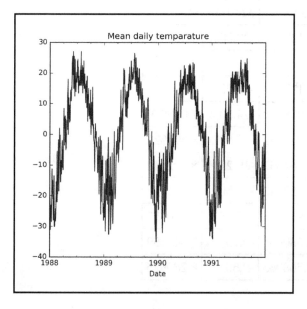

Figure 2.3: Time series of daily mean temperature of Fisher River in Texas, United States

The original time series seems to have monthly patterns that repeat every year and can be verified by calculating month wise averages. This is done by grouping-by the data into 12 months and then computing averages for each month. The code that does this is shown in the following snippet. We start by adding a `Month_Year` column to the `DataFrame`:

```
df['Month_Year'] = df.index.map(lambda d: d.strftime('%m-%Y'))
```

Next the `Mean temparature` column is grouped with respect to the newly added `Month_Year` column and month wise mean, median and standard deviations are computed:

```
monthly_stats = df.groupby(by='Month_Year')['Mean
temparature'].aggregate([np.mean, np.median,
np.std])
monthly_stats.reset_index(inplace=True)
monthly_stats.head(10)
```

The month wise aggregates are shown in the following table:

	Month_Year	mean	Median	std
0	01-1988	-22.137097	-23.00	5.260640
1	01-1989	-17.129032	-18.00	8.250725
2	01-1990	-15.112903	-12.00	6.606764
3	01-1991	-23.038710	-24.50	7.095570
4	02-1988	-19.025862	-19.50	8.598522
5	02-1989	-19.267857	-19.25	8.092042
6	02-1990	-17.482143	-16.50	8.018477
7	02-1991	-10.967857	-12.15	8.220753
8	03-1988	-8.258065	-9.25	5.341459
9	03-1989	-12.508065	-9.50	8.289925

Note that the rows in the preceding table are not in the ascending order of `Month_Year`. Therefore, it needs to be recorded. This is done by creating two new columns - `Month` and `Year` and then sorting in the ascending order of `Year` followed by sorting in the ascending order of `Month`:

```
monthly_stats['Year'] = monthly_stats['Month_Year']\
                                    .map(lambda m: pd.to_datetime(m,
format='%m-%Y').strftime('%Y'))
monthly_stats['Month'] = monthly_stats['Month_Year']\
                         .map(lambda m: pd.to_datetime(m, format='%m-
%Y').strftime('%m'))
monthly_stats.sort_values(by=['Year', 'Month'], inplace=True)
monthly_stats.head(10)
```

The first 10 rows of the sorted `DataFrame` are given here:

	Month_Year	mean	median	std	Year	Month
0	01-1988	-22.137097	-23.000	5.260640	1988	01
4	02-1988	-19.025862	-19.500	8.598522	1988	02
8	03-1988	-8.258065	-9.250	5.341459	1988	03
12	04-1988	2.641667	1.875	5.057720	1988	04
16	05-1988	11.290323	11.000	6.254364	1988	05
20	06-1988	19.291667	19.000	3.909032	1988	06
24	07-1988	19.048387	18.500	3.073692	1988	07
28	08-1988	17.379032	18.000	3.183205	1988	08
32	09-1988	10.675000	10.750	3.880294	1988	09
36	10-1988	2.467742	3.000	6.697245	1988	10

The monthly aggregates are plotted in the following figure, which highlights the month wise seasonality that exists in the original data.

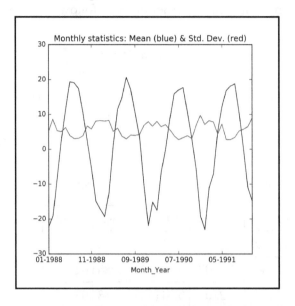

Figure 2.4: Monthly mean and standard deviation of temperature of Fisher River in Texas, United States

Moving statistics

In this section, we will continue working on the Fisher River dataset. Moving or rolling statistics are computed by sliding a window of certain size over the original time series and aggregating the data for every window. This is also known as **convolving** over the time index. The two important parameters of the **convolving** operation are window size and stride length. The former defines the number of time units taken as input to the aggregate function while the latter sets the gap along the time index between every computation. For example, suppose a window size of k and stride of length l is used to compute the function f over a time series $x_1, x_2, \ldots, x_n$ having N observations. In this scenario, the moving statistic is obtained as $f(x_1, x_2, \ldots, x_t)$, $f(x_{1+l}, x_{2+l}, \ldots, x_{t+l})$, and so on. Notice that every time the function is calculated by shifting the time window to the right by l time units.

Moving average is a special case of the function f and requires simply averaging the observations in the time window.

Let us demonstrate how moving averages can be computed on the Fisher River dataset. We will compute weekly moving averages that would set the window size to seven and slide the window by one place to the right:

```
weekly_moving_average = df['Mean temparature'].rolling(7).mean()
Calculating monthly averages can be done as follows with a window of size
thirty.
monthly_moving_average = df['Mean temparature'].rolling(30).mean()
```

The `rolling` function takes only the window size as argument. Therefore, to add a stride length of more than one we still calculate the moving average as shown previously, but then slice the resulting series to get the desired output. For strides over two time units we use the following code:

```
weekly_moving_average_2stride = df['Mean
temparature'].rolling(7).mean()[::2]
monthly_moving_average_2stride = df['Mean
temparature'].rolling(30).mean()[::2]
```

In time series analysis, moving statistics based on stride length of one is most common so you should rarely need anything more than the rolling function. The original data is plotted along with the weekly and monthly averages in the following figure, which shows the noise reduction and subsequent smoothing effect produced by the moving averages:

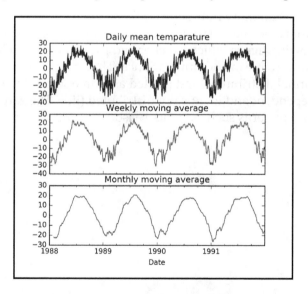

Figure 2.5: Moving averages of temperature of Fisher River in Texas, United States

Stationary processes

Properties of data such as central tendency, dispersion, skewness, and kurtosis are called sample statistics. Mean and variance are two of the most commonly used sample statistics. In any analysis, data is collected by gathering information from a sample of the larger population. Mean, variance, and other properties are then estimated based on the sample data. Hence these are referred to as sample statistics.

An important assumption in statistical estimation theory is that, for sample statistics to be reliable, the population does not undergo any fundamental or systemic shifts over the individuals in the sample or over the time during which the data has been collected. This assumption ensures that sample statistics do not alter and will hold for entities that are outside the sample used for their estimation.

This assumption also applies to time series analysis so that mean, variance and auto-correlation estimated from the simple can be used as a reasonable estimate for future occurrences. In time series analysis, this assumption is known as **stationarity**, which requires that the internal structures of the series do not change over time. Therefore, stationarity requires mean, variance, and autocorrelation to be invariant with respect to the actual time of observation. Another way of understanding stationarity is that the series has constant mean and constant variance without any predictable and repetitive patterns.

A popular example of a stationary time series is the zero-mean series which is a collection of samples generated from a normal distribution with mean at zero. The zero-mean series is illustrated in the following figure which is generated from points which are sampled from a normal distribution of zero mean and unit variance. Though points are sequentially sampled from the normal distribution and plotted as a time series, the individual observations are independent and identically distributed (iid). The zero-mean series does not show any temporal patterns such as trend, seasonality and auto-correlation.

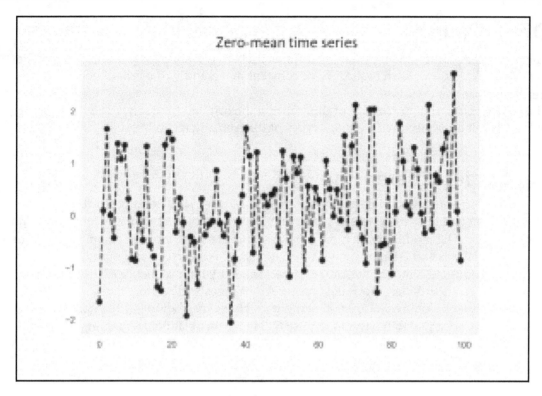

Figure 2.6: Zero-mean time series

However, most real-life time series are not stationary. Non-stationarity mostly arises due to the presence of trend and seasonality that affects the mean, variance, and autocorrelation at different points in time. However, it is noteworthy that we have not included cyclical fluctuations while defining stationarity. This is because crests and troughs due to cyclical changes do not occur at fixed intervals and therefore can be explained only by exogenous variables. In general, a time series with no predictable patterns in the long run (without considering exogenous factors as explanatory variables of course!) is stationary.

An crucial step in time series analysis is statistically verifying stationarity and destationarizing a non-stationary time series through special mathematical operations. To this end, we will discuss the **Augmented Dickey-Fuller** (**ADF**) test for detecting stationarity and describe the method of differencing for destationarizing non-stationary time series. Differencing can remove trend and seasonal components. The methods of decomposition to develop models of trend and seasonality for complex time series, is discussed in the next section.

Differencing

The basic idea of differencing is taking differences between successive occurrences of the time series $\Delta x_t = x_t - x_{t-1}$ such that Δx_t have constant mean and variance and hence can be treated as a stationary series.

The ADF test is based on the idea of differencing the original time series and therefore we will explain it after discussing various types of differencing techniques.

First-order differencing

First order differencing implies taking differences between successive realizations of the time series so that the differences Δx_t are irregular variations free from any long run trend or seasonality. The random walk model discussed in the last chapter is a sum of subsequent random variations and is given by $x_t = x_{t-1} + \epsilon_t$ where ϵ_t is a zero mean random number from normal distribution. Random walks are characterized by long sequence of upward or downward trends. Besides, they take unforeseen changes in direction. Based on these characteristics, random walks are non-stationary. However, the first differences (Δx_t of a random walk are equal to the random noise ϵ_t. Hence the residuals remaining after first-order differencing of a random walk is a zero-mean stationary series. The transformed time series, obtained by taking first order differences is denoted as follows:

$$x'_t = x_t - x_{t-1}$$

The transformed time series has *N-1* observations with zero mean and constant variance. This model assumes the starting value is $x_1 = 0$. The starting value can be generalized to $x_1 = 0$ and hence the the differenced time series would be:

$$x'_t = x_t - x_{t-1} = c + \epsilon t$$

If the starting value $x_1 = 0$ is positive, the random walk tends to drift upwards while if it is negative, the drift goes downward.

As evident from the preceding discussion that the first-order differences are independent and identically distributed with a constant mean and a constant variance and hence have no autocorrelation.

A quick way to verify whether the first-order differencing has stationarized a time series is to plot the ACF function and run the Ljung-Box test for the differenced series. The Ljung-Box test determines if the observed auto-correlation is statistically significant. The null hypothesis of the Ljung-Box test is that the time series consist of random variations and lacks predictable autocorrelation while the alternate hypothesis proposes that the observed autocorrelation is not random.

Let us illustrate the Ljung-Box test using the time series of **Dow Jones Industrial Average (DJIA)**, which was used in the last chapter to illustrate **autocorrelation function (ACF)**. We start by taking first difference of the DJIA Close values and plotting the resulting time series of first-order differences in the following figure:

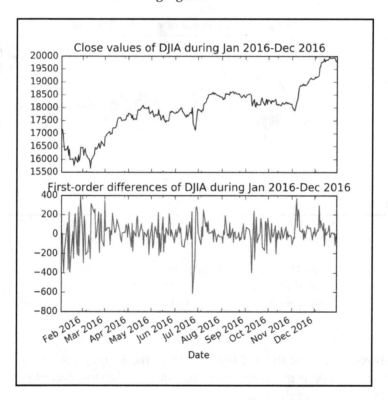

Figure 2.7: Time series of Dow Jones Industrial Average and its first-order differences

Next, the ACF is computed for different lags and verified by the Ljung-Box test. The following figure shows ACF of the DJIA close values and as well as for the time series of first-order differences. Notice that for the differenced series, ACF shows no predictable pattern and drops suddenly close to zero. Moreover, the p-value of the test being 0.894 for *lag=10* makes us accept the null hypothesis of Ljung-Box test for the differenced series.

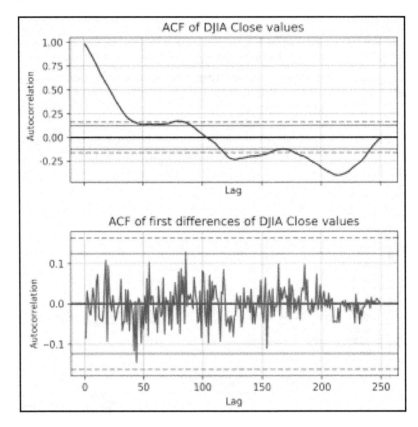

Figure 2.8: Autocorrelation of Dow Jones Industrial Average and its first-order differences

Now we will briefly discuss the code used for running the Ljung-Box test for the DJIA time series. Given that the original time series has been read into the `pandas.DataFrame` `djia_df`, the first-order differences are calculated as follows:

```
first_order_diff = djia_df['Close'].diff(1)
```

For the Ljung-Box test, we use the `acf` function from the `statsmodels.tsa.stattools` package. The `acf` function is made to return the autocorrelations, confidence intervals, Q-statistics, and the p-values of the test.

 The full code of this example is in the Jupyter notebook
`Chapter_2_First_Order_Differencing.ipynb` under the `code` folder of this book's GitHub repo.

Second-order differencing

In some cases, first-order differencing does not stationarize the time series and therefore the data is differenced another time to generate a stationary time series. Therefore, the second-order differenced time series is generated as follows:

$$x''_t = x'_t - x'_{t-1} = (x_t - x_{t-1}) - (x_{t-1} - x_{t-2}) = x_t - 2_{xt-1} + x_{t-2}$$

The time series resulting from second-order differencing have $N - 2$ observations. It is almost never required to perform differencing of order higher than second order.

Seasonal differencing

When a time series exhibits seasonality of a known time period of m time indices, it can be stationarized by taking seasonal differences between x_t and x_{t-m}. These differences taken at lags of length m are indicative of the seasons or quarters in a year. In this case $m = 12$ and the differences are taken between original observations that are a year apart from each other. The seasonal differences can be expressed as follows:

$$x'_t = x_t - x_{t-m} = \epsilon_t$$

To demonstrate the effect of seasonal differencing we would revisit the time series on daily mean temperatures of Fisher River. We have already seen the original time series and the monthly mean, both of which apparently exhibit strong seasonal behavior. The monthly mean can be obtained by running a month wise group-by operation that results in the pandas.Series object monthly_mean_temp. The autocorrelation in this series is computed and plotted using the autocorrelation_plot function from the pandas.plotting API and is shown in the following graph. The autocorrelation_plot function is useful to check the presence of statistically significant autocorrelation in a time series.

As shown in the following figure, it also plots the upper and lower confidence intervals for a confidence levels of 95% (alpha = 0.05; thick dotted line) and 99% (alpha = 0.01; thin dotted line). The ACF of the monthly mean temperature of Fisher River swings above and below the 99% confidence intervals for several lags. . The ACF of the monthly mean temperature of Fisher River swings above and below the 99% confidence intervals for several lags. Therefore, the monthly mean temperatures form a non-stationary time series due to seasonality.

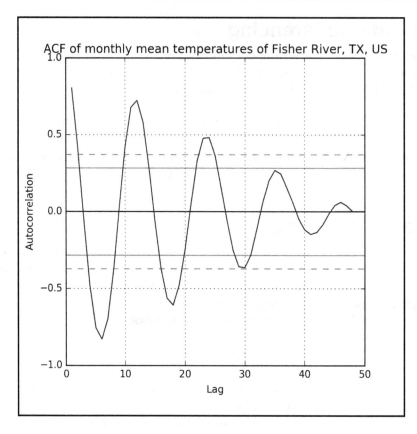

Figure 2.9: Autocorrelation of time series on monthly mean temperatures of Fisher River, Texas, United States

We attempt to stationarize the time series of monthly mean by taking seasonal differences as follows:

```
seasonal_diff = monthly_mean_temp.diff(12)
```

The preceding line of code returns `seasonal_diff`, which is a `pandas.Series`. The seasonal differencing leaves null values in its first 12 elements, which are removed before further analysis:

```
seasonal_diff = seasonal_diff[12:]
```

The seasonal differences appears to be random variations as shown in the following figure:

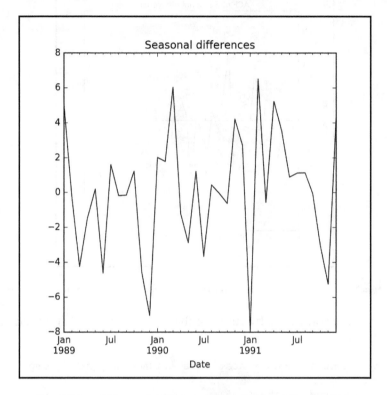

Figure 2.10: Seasonal differences of monthly mean temperatures of Fisher River, Texas, United States

We again use the `autocorrelation_plot` function to generate the ACF of the differenced series and the confidence intervals at 99% of confidence level. We can see in the following figure that the ACF never crosses the 99% confidence intervals for lags varying from 0 to over 35:

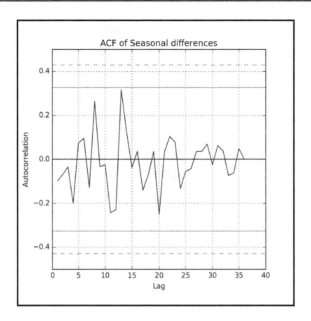

Figure 2.11: Autocorrelation of seasonal differences of monthly mean temperatures of Fisher River, Texas, United States

The actual p-values are confirmed by running the `stattools.acf` function on the monthly mean data as follows:

```
_, _, _, pval_monthly_mean = stattools.acf(monthly_mean_temp,
unbiased=True,
nlags=10, qstat=True, alpha=0.05)
print('Null hypothesis is rejected for lags:',
np.where(pval_monthly_mean<=0.05))
```

The preceding lines display the following message:

```
Null hypothesis is rejected for lags: (array([0, 1, 2, 3, 4, 5, 6, 7, 8,
9]),)
```

The Ljung-Box test is performed on the seasonally differenced series as well:

```
_, _, _, pval_seasonal_diff = stattools.acf(seasonal_diff, unbiased=True,
                                             nlags=10,
qstat=True, alpha=0.05)
print('Null hypothesis is rejected for lags:',
np.where(pval_seasonal_diff<=0.05))
```

The `print` function in the preceding code returns the following line:

```
Null hypothesis is rejected for lags: (array([], dtype=int32),)
```

There are no lags for which the null hypothesis of Ljung-Box's test is rejected.

 The full code for this example is in the Jupyter notebook `Chapter_2_Seasonal_Differencing.ipynb` under the `code` folder of this book's GitHub repo.

At this point, it is important to note that in some cases a first-order differencing is done after running the seasonal differences to achieve stationarity in the transformed data. The resulting series x''_t can be computed from the original as follows:

$$x''_t = x'_t - x'_{t-1} = (x_t - x_{t-m}) - (x_{t-1} - x_{t-m-1})$$

The choice of a differencing strategy can be determined through exploratory data analysis like the ones described so far. However, when it is difficult to determine what transformations are required for stationarization, the ADF test is performed for a definitive guidance.

Augmented Dickey-Fuller test

The statistical tests for objectively determining whether differencing is required to stationarize a time series are known as unit root tests. There are several such tests of which we discuss the ADF test is one of the unit root tests, that is most commonly used for verifying non-stationarity in the original time series. According to the ADF test, in the presence of autocorrelation, the first-order differences x'_t of the original series can be expressed as a linear regression model of the previous time index and first-order differences up to a lag of m times indices. The linear regression on x'_t can be formulated as follows:

$$x'_t = \gamma x_{t-1} + \beta_1 x'_{t-1} + \beta_2 x'_{t-2} + \cdots + \beta_m x'_{t-m} + \epsilon_t$$

In presence of strong autocorrelation, the original series needs differencing and hence the Y^x_{t-1} term in the preceding model would not add relevant information in predicting the change from time index $t-1$ and t. Therefore, the null hypothesis of the ADF is $H_{0:\gamma} = 0$ against the alternative hypothesis $H1_{.\gamma} < 0$. In other words, the null hypothesis is presence of unit root or non-stationarity, whereas the alternate hypothesis suggests stationarity of the data.

The null hypothesis is rejected of the test statistic for Y is less than the critical negative value for the given level of confidence. In case the series is already stationary, the linear regression model represents a random drift for which the change at time index t depends on the previous value and not the previous differences, which are iids for the stationary process.

The number of lags m to be included in the regression is usually set to three under the assumption that differencing of the order higher than third order differencing would be rarely needed to stationarize a time series. A practical approach of determining the m would be to vary $m \in \{1, 2, \dots, N'\}$, where N' is an upper limit on the lag, and compute p-values of the ADF test until the null hypothesis is rejected at a given level of confidence.

Let us apply the ADF to verify stationarity in the US Airlines monthly aircraft miles data collected during 1963 to 1970. We will use the `adfuller` function from the `statsmodels.tsa.stattools` API to run the tests.

Before running the ADF test, the time series is loaded into a `pandas.DataFrame` and plotted as shown in the following figure. It is evident that the time series has an uptrend as well as seasonality and therefore is non-stationary, which will be verified by the ADF test.

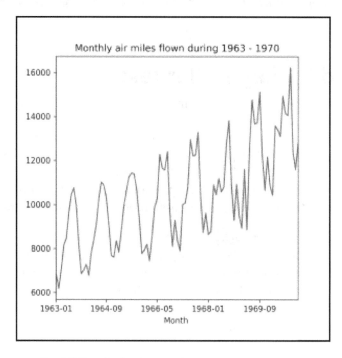

Figure 2.12: Time series of monthly air miles flown in United States during 1963-1969

The code for reading and plotting the time series is as follows:

```
air_miles = pd.read_csv('datasets/us-airlines-monthly-aircraft-miles-
flown.csv')
air_miles['Air miles flown'].plot(ax=ax)
```

Notice that we have used the built-in plot function of the `'Air miles flown'` column of the `air_miles`DataFrame. Now we are in a position to execute the ADF test using the `adfuller` function as follows:

```
adf_result = stattools.adfuller(air_miles['Air miles flown'],
autolag='AIC')
```

The first argument to the function is the second column of the DataFrame and the keyword argument `autolag='AIC'` instructs the function to choose a suitable number of lags for the test by maximizing the **Akaike Information Criteria** (**AIC**). Alternately, the test can run on the number of lags given by the user into the keyword argument maxlag. We prefer using the AIC over giving a lag to avoid trial and error in finding the best lag required for running the test.

Unsurprisingly, the p-value of the ADF test is 0.9945, which confirms our understanding from the exploratory data analysis. The `adfuller` function returns several values in a tuple. We have used the `adf_result` variable to store the results and refer to its second element `adf_result[1]` to retrieve the p-value of the test. Other interesting things returned by the function are `usedlag`, which is the number of lags actually used for running the test and critical values of the test statistic at 1%, 5%, and 10% levels of confidence.

Refer to the
`code/Chapter_2_Augmented_Dickey_Fuller_Test.ipynb` to see the
full code of this example.

Time series decomposition

The objective of time series decomposition is to model the long-term trend and seasonality and estimate the overall time series as a combination of them. Two popular models for time series decomposition are:

- Additive model
- Multiplicative model

The additive model formulates the original time series (x_t) as the sum of the trend cycle (F_t) and seasonal (S_t) components as follows:

$$x_t = F_t + S_t + \epsilon_t$$

The residuals ϵ_t obtained after adjusting the trend and seasonal components are the irregular variations. The additive model is usually applied when there is a time-dependent trend cycle component, but independent seasonality that does not change over time.

The multiplicative decomposition model, which gives the time series as product of the trend, seasonal, and irregular components is useful when there is time-varying seasonality:

$$xt = F_t \times S_t \times \epsilon_t$$

By taking logarithm, the multiplicative model is converted to an additive model of logarithm of the individual components. The multiplicative model is expressed as follows:

$$\log(x_t) = \log(F_t) + \log(S_t) + \log(\epsilon_t)$$

In this section, we will discuss the following two popular methods for estimating the trend and seasonal components:

- Method of Moving Averages
- Seasonal and Trend Decomposition using the Python package `statsmodels.tsa`

Moving averages

In this section, we will cover moving averages through the following topics:

- Computation of moving averages and their application in smoothing time series
- Seasonality adjustment using moving average
- Weighted moving average
- Time series decomposition using moving averages

Moving averages and their smoothing effect

Moving averages (**MA**) at a time index t estimates the average trend cycle component F_t and is calculated by taking average of over the time period of $t \pm k$ where k is the range of the MA:

$$\hat{F}_t = \frac{x_{t-k} + x_{t-k+1} + \cdots + x_t + \cdots + x_{t+k-1} + x_{t+k}}{2k + 1}$$

Taking moving averages have an effect of smoothing the original time series by eliminating random noise. Commonly the total number of observations $m = 2k + 1$ is used to describe the moving average as m-order MA, which henceforth will be denoted as $\hat{F}_t^{(m)}$. Let us demonstrate moving averages and their smoothing effect through the example on IBM stock prices from 1962 to 1965. The dataset for this time series is downloaded from `http://datamarket.com`. The original time series, as shown in blue in the following figure, has irregular movements due to random noise. The 5 day (or 5- order) MA, displayed in red, is smoother than the original series and shows an estimate of the trend-cycle pattern. The 5-day MA has apparently produced some smoothing effect on the original time series as seen in the following figure.

> The code for this example is in the notebook
> `Chapter_2_Moving_Averages.ipynb` under the code folder of this book's GitHub repo.

The original time series has been loaded into a `panda.DataFrame`. Then, the second column, having the stock prices, is renamed to `Close_Price` for convenience. Next, the 5-day MA is computed by applying the rolling function on the `Close_Price` column of the `DataFrame`. The implementation is as follows:

```
ibm_df = pd.read_csv('datasets/ibm-common-stock-closing-prices.csv')
ibm_df.rename(columns={'IBM common stock closing prices': 'Close_Price'},
        inplace=True)
```

```
ibm_df['5-Day Moving Avg'] = ibm_df['Close_Price'].rolling(5).mean()
The plotting is done using the in-built plot function of pandas.Series
objects.
fig = plt.figure(figsize=(5.5, 5.5))
ax = fig.add_subplot(2,1,1)
ibm_df['Close_Price'].plot(ax=ax)
ax.set_title('IBM Common Stock Close Prices during 1962-1965')
ax = fig.add_subplot(2,1,2)
ibm_df['5-Day Moving Avg'].plot(ax=ax, color='r')
ax.set_title('5-day Moving Average')
```

Figure 2.13: IBM stock price and its 5-day moving average during 1962-1965

The aforementioned MA with an odd order $m = 2 \times 2 + 1 = 5$ and is symmetric with equal number of observations on both sides of the time index t at which the MA is being calculated. However, it is possible to have an asymmetric MA of even order $m = 2k$.

The asymmetry of an even order moving average can be eliminated by taking a second moving average of an even order. Let us illustrate this by considering the first moving average to be of the order two and computed as:

$$\hat{F}_t^{(2)} = \frac{x_{t-1} + x_t}{2}$$

Another second order moving average, when applied on the series $\hat{F}_t^{(2)}$, will produce a final symmetric moving average that has one observation on both sides of the observation from the original time series at index t:

$$2 \times \hat{F}_t^{(2)} = \frac{\hat{F}_t^{(2)} + \hat{F}_{t+1}^{(2)}}{2} = \frac{1}{2}\left[\frac{x_{t-1} + x_t}{2} + \frac{x_t + x_{t+1}}{2}\right] = \frac{1}{4}x_{t-1} + \frac{1}{2}x_t + \frac{1}{4}x_{t+1}$$

In general, we can create $n \times \hat{F}_t^{(m)}$ moving averages by first taking an m order moving average followed by an n order moving average. However, we must note that in order to produce symmetric MA, m and n should either be both even or both odd. We will clarify this through the computations of a $3 \times \hat{F}_t^{(3)}$ MA, which is given as follows:

$$\hat{F}_t^{(3)} = \frac{x_{t-1} + x_t + x_{t+1}}{3}$$

$$3 \times \hat{F}_t^{(3)} = \frac{\hat{F}_{t-1}^{(3)} + \hat{F}_t^{(3)} + \hat{F}_{t+1}^{(3)}}{3}$$

The $3 \times \hat{F}_t^{(3)}$ MA can be further broken down in terms of the original series as follows:

$$3 \times \hat{F}_t^{(3)} = \frac{1}{3}\left[\frac{x_{t-2} + x_{t-1} + x_t}{3} + \frac{x_{t-1} + x_t + x_{t+1}}{3} + \frac{x_t + x_{t+1} + x_{t+2}}{3}\right]$$
$$= \frac{1}{9}x_{t-2} + \frac{2}{9}x_{t-1} + x_t + \frac{2}{9}x_{t+1} + \frac{1}{9}x_{t+2}$$

In contrast let us consider a $2 \times \hat{F}_t^{(3)}$ that can be expressed in terms of the original time series as follows:

$$2 \times \hat{F}_t^{(3)} = \frac{1}{2}\left[\frac{x_{t-1} + x_t + x_{t+1}}{3} + \frac{x_t + x_{t+1} + x_{t+2}}{3}\right] = \frac{1}{6}x_{t-1} + \frac{1}{3}x_t + \frac{1}{3}x_{t+1} + \frac{1}{6}x_{t+2}$$

which has more number of observations on the right of x_t than on the left and therefore it is asymmetric. As it would turn, symmetric MAs have interesting application of estimating trend-cycle for seasonal time series. This is explained in the next subsection.

It is also noteworthy that a repeated moving average of the form $n \times \hat{F}_t^{(m)}$ produces weighted averages of observations from the original time series. The weight assigned to observation at time index t is highest and drops as we move further away from t on both sides. Hence, $n \times \hat{F}_t^{(m)}$ can be used to generate weighted moving averages.

The plot given here shows six moving averages namely $\hat{F}_t^{(2)}$, $2 \times \hat{F}_t^{(2)}$, $\hat{F}_t^{(4)}$, $2 \times \hat{F}_t^{(4)}$, $F_t^{(3)}$, and $3 \times \hat{F}_t^{(3)}$ for IBM stock prices for up to the first 45 days. As illustrated, the smoothness of the resulting MAs improves with increase in the order m and number of repetitions n.

In the following figure, MAs of type $\hat{F}_t^{(m)}$ are given in solid lines while those of type $n \times \hat{F}_t^{(m)}$ are plotted in dashed lines.

To generate the six aforementioned MAs we have extensively used the `rolling` and `mean` functions as shown in the following code snippets:

```
MA2 = ibm_df['Close_Price'].rolling(window=2).mean()
TwoXMA2 = MA2.rolling(window=2).mean()

MA4 = ibm_df['Close_Price'].rolling(window=4).mean()
TwoXMA4 = MA4.rolling(window=2).mean()

MA3 = ibm_df['Close_Price'].rolling(window=3).mean()
ThreeXMA3 = MA3.rolling(window=3).mean()
```

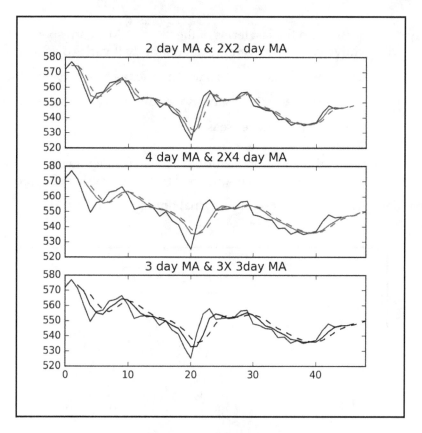

Figure 2.14: Different moving averages of IBM stock price

Seasonal adjustment using moving average

The weighted averaging property of the $n \times \hat{F}_t^{(m)}$ moving averages has application in smoothing data with seasonality in order to generate estimates of trend cycles. For example, given quarterly observations we can apply the $2 \times \hat{F}_t^{(4)}$ MA to smooth the data. To understand how this works let us expand $2 \times \hat{F}_t^{(4)}$ in terms of the original time series:

$$2 \times \hat{F}_t^{(4)} = \frac{\hat{F}_t^{(4)} + \hat{F}_{t+1}^{(4)}}{2} = \frac{1}{2}\left[\frac{x_{t-2} + x_{t-1} + x_t + x_{t+1}}{4} + \frac{x_{t-1} + x_t + x_{t+1} + x_{t+2}}{4}\right]$$

$$= \frac{x_{t-2}}{8} + \frac{x_{t-1}}{4} + \frac{x_t}{4} + \frac{x_{t+1}}{4} + \frac{x_{t+2}}{8}$$

For quarterly data, the first and the last terms of the $2 \times \hat{F}_t^{(4)}$ MA corresponds to the same quarter, but in consecutive years and averages out quarterly it varies over a year. Similarly, for monthly data, $2 \times \hat{F}_t^{(12)}$ would generate a smooth series as an estimate of the trend-cycle component. For data with odd periodicity such as weekly observations, taking $\hat{F}_t^{(2k+1)}$ MA, for example $\hat{F}_t^{(7)}$ for weekly data removes seasonality.

The approach of using weighted MA of the form $2 \times \hat{F}_t^{(2k)}$ is illustrated using the beer production data from Australia. The time series represents quarterly beer production is taken during March 1956 to June 1994. The original time series and a series of $2 \times \hat{F}_t^{(4)}$ MA are plotted here:

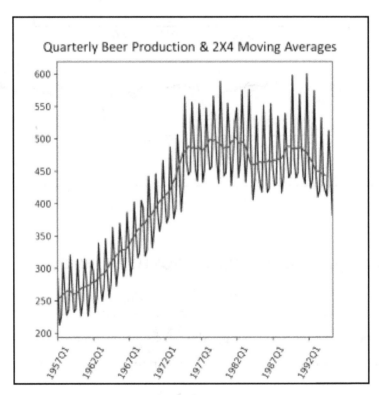

Figure 2.15: Quarterly beer production and its 2 by 4 moving average during 1957-1992

The code to read the dataset and compute the moving averages is as follows:

```
beer_df = pd.read_csv('datasets/quarterly-beer-production-in-aus-March
1956-    June 1994.csv')
beer_df.rename(columns={'Quarterly beer production in Australia:
megalitres.   March 1956 ? June 1994': 'Beer_Prod'},
            inplace=True
            )
MA4 = beer_df['Beer_Prod'].rolling(window=4).mean()
TwoXMA4 = MA4.rolling(window=2).mean()
TwoXMA4 = TwoXMA4.loc[~pd.isnull(TwoXMA4)]
```

The original time series on the quarterly beer productions has trend as well as seasonality and therefore is not stationary. Let us see if we can stationarize the time series by first removing the trend component and then taking seasonal differences.

We start by taking residuals left after removing the trend component:

```
residuals = beer_df['Beer_Prod']-TwoXMA4
residuals = residuals.loc[~pd.isnull(residuals)]
```

The `residuals` left after removing the trend-cycle component are plotted in the following figure:

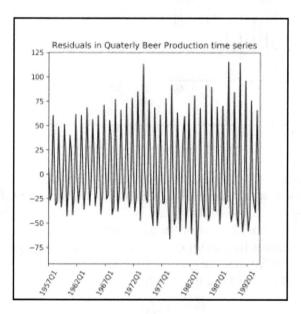

Figure 2.16: Residuals remaining after taking the difference of quarterly beer production and its moving average

At this point we will check whether the residuals are already stationarized (though unlikely!) by plotting the autocorrelation function along with the 99% confidence intervals. To get this plot, which is shown here, we would make use of the `autocorrelation_plot` function from the `pandas.plotting` API:

```
autocorrelation_plot(residuals, ax=ax)
```

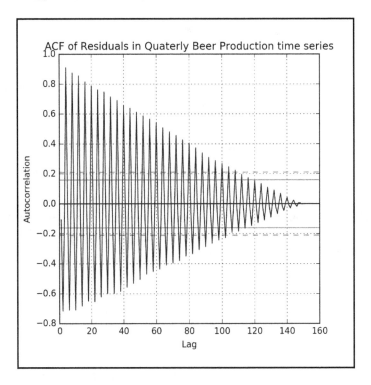

Figure 2.17: Autocorrelation of quarterly beer production

Clearly, the residuals have a strong autocorrelation with the ACF jumping outside the confidence intervals for several values of lags. So we would need to take seasonal difference on the residuals. The period of seasonality can be determined based on the fact that the original data is obtained from all quarters of the years and shows seasonality of the quarter. This means that the residuals in quarter one of a year is close in magnitude to the residuals from quarter one of the preceding and succeeding years. This observation makes us take differences over periods of four time units as follows:

```
residuals_qtr_diff = residuals.diff(4)
residuals_qtr_diff = residuals_qtr_diff.loc[~pd.isnull(residuals_qtr_diff)]
```

We expect the `residuals_qtr_diff` to be a series of random variations with no seasonality and predictable autocorrelation. To verify if this is the case, we run the `autocorrelation_plot` function on `residuals_qtr_diff` and obtain the following graph that has mostly randomly ACF. Besides, the ACF falls outside the 99% confidence intervals only for two lags. This means that taking seasonal differences has stationarized the residuals.

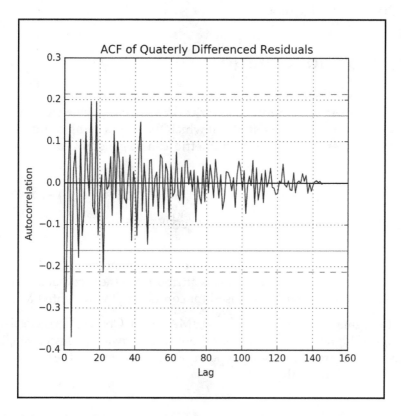

Figure 2.18: Autocorrelation of residuals remaining after taking the difference of quarterly beer production and its moving average

Our approach for stationarizing the quarterly beer production time series can be summarized as follows:

1. Take seasonal $2 \times \hat{F}_t^{(4)}$ MA.

2. Compute residuals by removing the $2 \times \hat{F}_t^{(4)}$ MA from the original time series.

3. Check for randomness of the ACF of the residuals.

4. If the residuals' ACF is already random then the residuals are stationary, if not proceed to the next step.
5. Take seasonal differences on the residuals with periodicity of four and check for randomness of the ACF of the differenced series.

Hence, seasonal moving averages and seasonal differences on the residuals have essentially stationarized the original time series.

Weighted moving average

In the previous sections, we expressed moving averages of the form $n \times \hat{F}_t^{(m)}$ as weighted sum of observations from the original time series. We saw how the weights drop for observations far away from the time index at which the MA is being computed. The notion of weighted moving averages, which is symmetric, can be generalized for any time series application as follows:

$$\hat{F}_t = w_{t-k}x_{t-k} + w_{t-k+1}x_{t-k+1} + \cdots + w_t x_t + \cdots w_{t+k-1}x_{t+k-1} + w_{t+k}x_{t+k}$$

Where the weights $w_{t-k} + w_{t-k+1} + \cdots + w_{t+k-1} + w_{t+k} = 1$

For a simple $\hat{F}_t^{(m)}$ the weights are $\frac{1}{m}$. It has been observed that weighted MA yields smoother estimation of the trend cycle. This is attributed to the fact that all observation from the original time series, which are used for computing the weighted MA are not assigned equal weights as in case of a $n \times \hat{F}_t^{(m)}$ MA series. Observations close to the time index t are assigned higher weight, whereas those farther away get lesser weights. This approach of assigned different weights depending on proximity to the time index of interest generates better smoothing of the data.

Time series decomposition using moving averages

While analysing the quarterly beer production data we developed a methodology of stationarizing a non-stationary time series using seasonal MA and seasonal differences. We used the seasonal moving average as an estimate of the trend-cycle component and computed periodic differences on the residuals left by the MA.

Another approach could have been deducting both the seasonal MA and seasonal residuals from the original series and checking for randomness of the final residuals. This approach assumes that the beer production series is an additive sum of the trend-cycle and seasonal components and what is left after removing the aforementioned two are random variations. Indeed, moving averages can be used in this manner to decompose a time series. Let us elaborate the moving average driven time series decomposition in this section.

Moving averages of the form $n \times \hat{F}_t^{(m)}$ have the property of smoothing the original time series and giving an estimation of the trend cycle component. Choice of m and n are crucial for determining the trend-cycle component. Commonly, m is the periodicity of the seasonal data and is known apriori or determined through exploratory data analysis. If m is even, $\hat{F}_t^{(m)}$ moving averages would be asymmetric and therefore lack the necessary smoothing properties over successive occurrences of the seasons. Hence, moving averages are taken for the second time to generate $2 \times \hat{F}_t^{(m)}$ series which has the necessary seasonal smoothing and gives a better estimate of the trend-cycle component. However in case of odd periodicity, $\hat{F}_t^{(m)}$ is used as the estimate of the trend-cycle component. Therefore, estimate of the trend cycle component is $\hat{F}_t = 2 \times \hat{F}_t^{(m)}$ or $\hat{F}_t^{(m)}$ depending on whether m is even or odd.

In time series decomposition using moving averages, the seasonal component is assumed to be constant from a year to year or from a week to week. Estimate of the seasonal component is generated by taking season-wise average of the residuals left after trend-cycle adjustment. For example, if the data is monthly, we take month wise average of the residuals from trend-cycle adjustment.

The computations will slightly differ between an additive and multiplicative model. For example, upon estimation of the trend-cycle component $\hat{F}_t$ by application of a suitable moving average, the residuals for an additive model is calculated by $x_t - \hat{F}_t$, whereas for a multiplicative model, residuals are $\frac{x_t}{\hat{F}_t}$. The seasonal component $\hat{S}_t$ is now estimated from the residuals as season wise averages.

Lastly, the irregular variations are obtained by making trend-cycle and seasonal adjustment of the original series as follows:

- For an additive decomposition model: $\epsilon_t = x_t - \hat{F}_t - \hat{S}_t$

- For a multiplicative decomposition model: $\epsilon_t = \frac{x_t}{\hat{F}_t \times \hat{S}_t}$

Let us illustrate this approach of time series decomposition using moving averages through the time series on monthly aircraft mils flown by US Airlines. We apply both additive and multiplicative models for this dataset.

We start by reading the dataset into a `pandas.DataFrameair_miles`. Next the trend-cycle component is estimated by $2 \times \hat{F}_t^{(12)}$ moving averages as follows:

```
MA12 = air_miles['Air miles flown'].rolling(window=12).mean()
trendComp = MA12.rolling(window=2).mean()
```

For the additive model, the seasonal component is obtained by subtracting the trend-cycle from the original time series and taking month wise averages for the residuals. The group-by operation is used to compute the seasonal component as follows:

```
residuals = air_miles['Air miles flown'] - trendComp

#To find the sesonal compute we have to take monthwise average of these
residuals
month = air_miles['Month'].map(lambda d: d[-2:])
monthwise_avg = residuals.groupby(by=month).aggregate(['mean'])
#Number of years for which we have the data
nb_years = 1970-1963+1

seasonalComp =
np.array([monthwise_avg.as_matrix()]*nb_years).reshape((12*nb_years,))
```

Given the estimates of trend-cycle and seasonal components, we get the irregular variations as follows for an additive model:

```
irr_var = air_miles['Air miles flown'] - trendComp - seasonalComp
```

The original time series, the trend-cycle, seasonal, and irregular components are shown in the following figure:

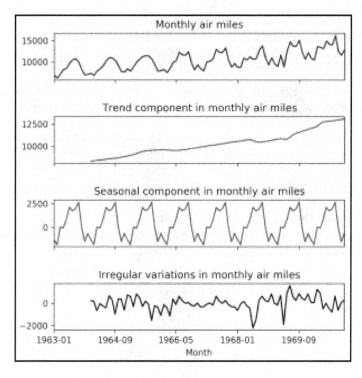

Figure 2.19: Additive decomposition of time series on monthly air miles

ADF test on the `irr_var` gives a p-value of 0.0658. At confidence level of 90% (alpha=0.10), the null hypothesis on stationarity of the irregular variations can be accepted. However, let us try for further improvement through the multiplicative model.

The trend-cycle estimation by $2x\widehat{F}_t^{(12)}$ MA is applicable even for the multiplicative model. However, calculation of the seasonal component changes:

```
residuals = air_miles['Air miles flown'] / trendComp
```

Notice that the seasonal `residuals` are dividing the trend-cycle component from the original time series in case of a multiplicative model. However, the group-by operation on the residuals remain the same.

Finally, we reach at the irregular variations for the multiplicative model as follows:

```
irr_var = air_miles['Air miles flown'] / (trendComp * seasonalComp)
```

ADF test on irregular variations obtained from the multiplicative model gives a p-value of approximately 0.00018, which is much lesser than that obtained from the additive model.

 The reader is encouraged to have a look at the full code of the analysis done in this section in the Jupyter notebook `code/Chapter_2_Time_Series_Decomposition_by_Moving_Averag es.ipynb`.

The original time series along with the trend-cycle, seasonal, and irregular components are plotted in the following figure. Notice that the seasonal and irregular components are much smaller in magnitude than those obtained from the additive model:

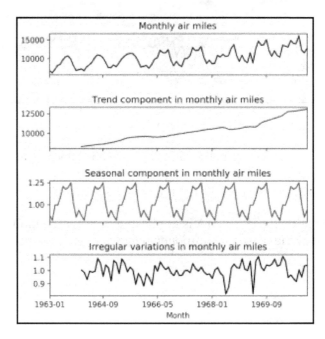

Figure 2.20: Multiplicative decomposition of time series on monthly air miles

Time series decomposition using statsmodels.tsa

So far, we have discussed how MA can be used for estimating the trend-cycle and seasonal components of a time series. The method of MA works under the simple assumption that seasonal changes are constant over consecutive years, weeks, or a period suitable for the given use case. However, constant seasonality might be valid for several applications that require advanced method such as Seasonality and Trend decomposition using Locally Weighted Smoothing of Scatter plot also commonly referred as the STL method.

In this section, we will tackle time series with complex patterns using the Python `statsmodels.tsa` package. Our objective would be to estimate the trend-cycle and seasonal components. Besides, we would use trend-cycle and seasonal estimates to stationarize the data that would be verified by the ADF test.

Let us consider the time series on monthly employment in Wisconsin, US to illustrate the aforementioned approach. The data is over the period Jan 1961 to October 1975 and has been downloaded from `http://datamarket.com`.

We start by loading the dataset into a `pandas.DataFramewisc_emp` and running the ADF test for the original time series as follows:

```
wisc_emp = pd.read_csv('datasets/wisconsin-employment-time-series.csv')
adf_result = stattools.adfuller(wisc_emp['Employment'], autolag='AIC')
```

A high p-value of 0.9810 for the ADF test on the monthly employment series indicates that the original time series is non-stationary. Hence we attempt to decompose the time series and consequently stationarize it by using the `seasonal.seasonal_decompose` function from the `statsmodels.tsa` API. Let us first attempt the additive model for decomposition:

```
decompose_model = seasonal.seasonal_decompose(wisc_emp.Employment.tolist(),
freq=12, model='additive')
```

The argument `freq` in the `seasonal.seasonal_decompose` is the periodicity of the seasonal behavior and the original time series being monthly observations we suspect a periodicity of 12, which can be verified through exploratory data analysis.

The trend-cycle, seasonal, and residual components of the decomposed time series are accessible through the attributes of the object `decompose_model` returned by the `seasonal.seasonal_decompose` function. These components are can be obtained from the following attributes of the decompose_model:

- `decompose_model.trend` - Trend-cycle component

- decompose_model.seasonal - **Seasonal component**
- decompose_model.resid - **Irregular variations**

Now we run the ADF test on the residuals of the additive model and obtain a p-value of 0.00656, which is much smaller than that obtained from the original time series. However, we will build a multiplicative model as well:

```
decompose_model = seasonal.seasonal_decompose(wisc_emp.Employment.tolist(),
freq=12, model='multiplicative')
```

ADF test on the residuals from the multiplicative decomposition gives a p-value of 0.00123 that is even lesser than that obtained from the additive decomposition. At a confidence level of 99% (alpha=0.01) we can reject the null hypothesis of the ADF test and conclude that the residuals of the multiplicative decomposition model is a stationary series.

 The code for running the examples in this section is in the Jupyter notebook
code/Chapter_2_Time_Series_Decomposition_using_statsmodel
s.ipynb.

The original time series is plotted with the individual components in the following figure:

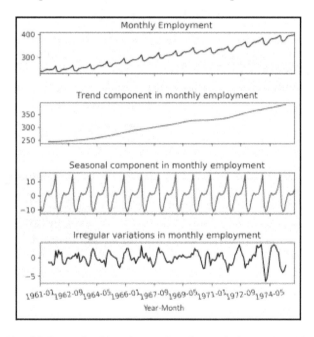

Figure 2.21: Decomposition of time series on monthly employment using the statsmodels.tsa API

Summary

We started this chapter by discussing advanced data processing techniques such as resampling, group-by, and moving window computations to obtain aggregate statistics from a time series. Next, we described stationary time series and discussed statistical tests of hypothesis such as Ljung-Box test and Augmented Dickey Fuller test to verify stationarity of a time series. Stationarizing non-stationary time series is important for time series forecasting. Therefore, we discussed two different approaches of stationarizing time series.

Firstly, the method of differencing, which covers first, second, and seasonal differencing, has been described for stationarizing a non-stationary time series. Secondly, time series decomposition using the `statsmodels.tsa` API for additive and multiplicative models has been discussed.

In the next chapter, we delve deeper in techniques of exponential smoothing which deals with noisy time series data.

3
Exponential Smoothing based Methods

The current chapter covers data smoothing on time series signal. The chapter is organized as follows:

- Introduction to time series smoothing
- First order exponential smoothing
- Second order exponential smoothing
- Modeling higher order exponential smoothing
- Summary

Introduction to time series smoothing

Time series data is composed of signals and noise, where signals capture intrinsic dynamics of the process; however, noise represents the unmodeled component of a signal. The intrinsic dynamics of a time series signal can be as simple as the mean of the process or it can be a complex functional form within observations, as represented here:

$$x_t = f(x_i) + \varepsilon_t \text{ for } i=1,2,3, \ldots t\text{-}1$$

Here, x_t is observations and ε_t is white noise. The $f(x_i)$ denotes the functional form; an example of a constant as a functional form is as follows:

$$x_t = \mu + \varepsilon_t$$

Here, the constant value **μ** in the preceding equation acts as a drift parameter, as shown in the following figure:

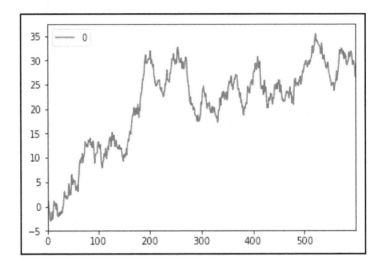

Figure 3.1: Example of time series with drift parameter

As ε_t is white noise, this smoothing-based approach helps separate the intrinsic functional form from random noise by canceling it. The smoothing forecasting methods can be considered as filters that take inputs and separate the trend and noise components, as shown in the following figure:

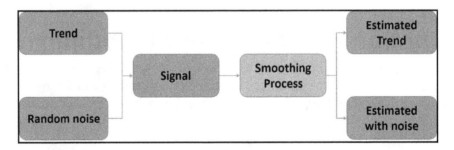

Figure 3.2: Framework for the smoothing process

The efficacy of the extraction of estimated trend and noise depends on other parameters related to time series signal composition such as the presence of trend, seasonality, and residual (noise). To handle each of these components of time series, different treatments are required. This chapter will cover multiple approaches to smooth to handle different components of the time series signal. An example of a time series signal composed of trend, seasonality, and white noise is demonstrated using the New York birth dataset collected at a monthly level from Jan 1946 till Dec 1959:

```
Import requests
import statsmodels.api as sm
import io
import pandas as pd

# Load Dataset
DATA_URL="http://robjhyndman.com/tsdldata/data/nybirths.dat"
fopen = requests.get(DATA_URL).content
ds=pd.read_csv(io.StringIO(fopen.decode('utf-8')),  header=None,
names=['birthcount'])
print(ds.head())

# Add time index
date=pd.date_range("1946-01-01", "1959-12-31", freq="1M")
ds['Date']=pd.DataFrame(date)
ds = ds.set_index('Date')

# decompose dataset
res = sm.tsa.seasonal_decompose(ds.birthcount, model="additive")
resplot = res.plot()
```

The different functions in the code are used as follows:

- The `requests.get` function in the preceding script is used to get the data from url DATA_URL.
- To handle the dataset, a pandas DataFrame is used.

- The `seasonal_decompose` function from the stats models module is utilized to decompose the time series signal into trend, seasonality, and residual components. The decomposition can be additive or multiplicative as discussed in `Chapter 2`, *Introduction to Time Series*. An example of different components of a signal is shown in the following figure:

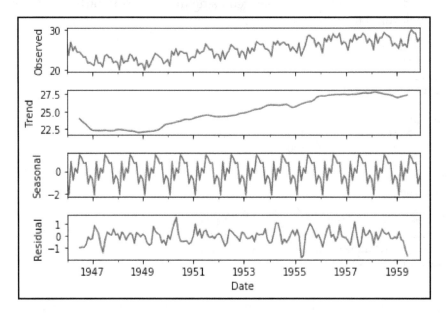

Figure 3.3: Decomposition of time series signal

The preceding time series signal is composed of trend, seasonality, and residual (noise). The smoothing helps remove the residual component, as shown in Figure *Framework for smoothing process*, and captures the trend and seasonality components for the forecasting signal.

The first step model to include mean, trend, and non-seasonal patterns is to extrapolate it using smoothing. The basic smoothing using moving average is discussed in `Chapter 2`, *Understanding Time Series data*. The moving average smoothing evaluates expectations *E(xt)* using all previous observations, as follows:

$$\hat{x}_t = \frac{1}{N} \sum_i x_i$$

Usually, simple moving average is performed on window; thus, the estimated forecast is evaluated on optimal windows with the objective function to minimize error:

$$\min\left(\sum (x_t - \hat{x}_t)^2\right)$$

$$\min \sum \left(x_t - \left(\frac{1}{N}\sum_t x_{t-1}\right)\right)^2$$

The smoothing methods are based on the assumption that time series data is locally stationary with small variations in the mean. Thus, we can use the mean at time *t* to predict at *t+1* where time delta is small enough to keep the signal stationary. This model is a compromise between mean and random walk without drift model. The models are also referred to as smoothing models as it smoothens out bumps from the dataset. The major limitation with these moving average-based methods is that it treats all *n* samples used in smoothing equally and ignores the observation recency effect, that is, giving higher weights to recent observations, as shown in the following equation:

$$\hat{x}_t = \frac{1}{N}\sum_i w_i x_i$$

Here, $w_1 > w_2 > w_3 > ... > w_T$ and T is window size. This places another challenge for the evaluation of weights. The limitation from moving average and weighted moving average is addressed by exponential smoothing by application of exponential decaying weights on observations.

First order exponential smoothing

First order exponential smoothing or simple exponential smoothing is suitable with constant variance and no seasonality. The approach is usually recommended to make short-term forecast. Chapter 2, *Understanding Time Series data,* has introduced the naïve method for the forecasting where prediction in horizon h is defined as value of t (or the last observation):

$$x_{t+h} = x_t$$

The approach is extended by simple moving average, which extends the naïve approach using the mean of multiple historical points:

$$X_{t+h} = \frac{1}{n}\sum_{i=1}^{n} x_t$$

The approach assumes equal weight to all historical observations, as shown in the following figure:

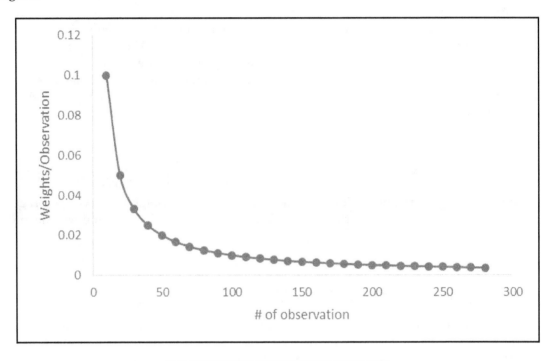

Figure 3.4: Weight assigned to observation with increasing window size

As the window size for moving average increases, the weights assigned to each observation become smaller. The first order exponential extends this current approach by providing exponential to historical data points where weights decrease exponentially from the most recent data point to the oldest. The first order exponential smoothing can be defined as follows:

$$X_{t+1} = \alpha x_t + \alpha(1-\alpha)x_{t-1} + \alpha(1-\alpha)^2 x_{t-1} + \dots$$

Here, α is the smoothing factor between [0,1] and controls the rate at which weights decrease and x_t is the observed value at time t. The following figure for the smoothing process demonstrates the decay of weights with a different smoothing factor, α:

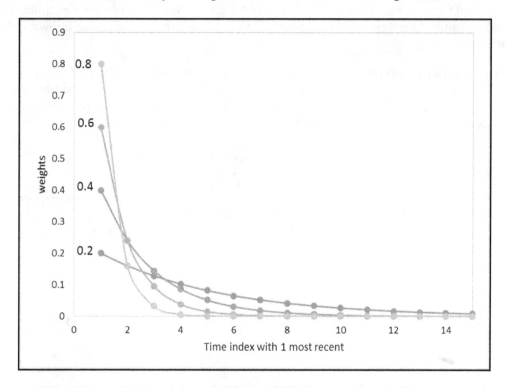

Figure 3.5: Illustration of weights assigned at different alpha values to historical observations

The higher value of α leads to a faster decay of weights; thus, historical data will have less impact on the forecasted value. The forecasting equation for single order exponential smoothing can be further reduced to the following:

$$F_1 = x_1$$

$$F_2 = \alpha x_1 + (1\text{-}\alpha)F_1 = \alpha x_1 + (1\text{-}\alpha)x_1 = x_1$$

$$F_3 = \alpha x_2 + (1\text{-}\alpha)F_2$$

$$...$$

$$F_{t\text{-}1} = \alpha x_t + (1\text{-}\alpha)F_{t\text{-}1}$$

Here, F_{t-1} is the forecasted value at time t and is represented as follows:

$$F_{t-1} = \alpha x_{t-2} + \alpha(1-\alpha)x_{t-3} + \alpha(1-\alpha)^2 x_{t-4} + \ldots$$

The reduced form demonstrates the relation of the forecasted value of time t as function forecasted value at time t-1 and observation at time t. The forecasting is also known as Holt-Winters forecasting.

As can be seen in the preceding equation, the initial value to the first forecast value is assigned an observation value in the first time instance, $F_1 = x_1$, and the second time instance also takes the value of the first instance. Another option to bootstrap the forecasting is to use the average value of the available data or a subset of the data, as follows:

$$F_1 = \frac{1}{N}\sum_{i=1}^{N} x_i$$

The analogous issue of detecting window size and weights in moving average exponential smoothing requires the optimization of the smoothing parameter α. The right selection of α is very critical. For example, selection of $\alpha=0$ will smoothen all values to the initial assigned value as shown:

$$F_{t+1} = F_{t-1}$$

Similarly, $\alpha=1$ represents the most unsmoothed case of exponential smoothing as follows:

$$F_{t+1} = x_t$$

Let's take an IBM common stock closing example to forecast using the single smoothing method. The first step is to load the required modules:

```
# Load modules
from __future__ import print_function
import os
import pandas as pd
import numpy as np
from matplotlib import pyplot as plt
```

The current example will use `os`, `pandas`, `numpy`, and `matplotlib` modules from Python. The dataset is loaded in the Python environment using `pandas DataFrame`:

```
# Load Dataset
ibm_df = pd.read_csv('datasets/ibm-common-stock-closing-prices.csv')
ibm_df.head()
```

Just for convenience, we will rename the columns:

```
#Rename the second column
ibm_df.rename(columns={'IBM common stock closing prices':
'Close_Price'},inplace=True)
```

The output dataset will look as shown in *Figure 3.6*:

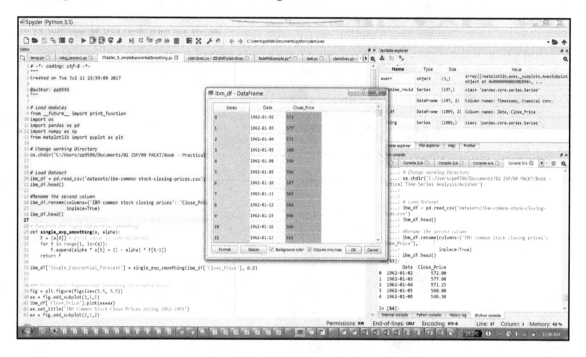

Figure 3.6: Sample IBM closing stock price dataset

In the single smoothing exponential method, the forecasted values are generated as follows:

$$\hat{x}_0 = x_0$$
$$\hat{x}_1 = \alpha x_0 + (1 - \alpha)\hat{x}_0$$

$$\ldots$$

$$x_{t+1} = \alpha x_t + \alpha(1 - \alpha)x_{t-1} + \alpha(1 - \alpha)^2 x_{t-1} + \ldots$$

The preceding series can be implemented in Python in the following way:

```python
# Function for Single exponential smoothing
def single_exp_smoothing(x, alpha):
    F = [x[0]] # first value is same as series
    for t in range(1, len(x)):
        F.append(alpha * x[t] + (1 - alpha) * F[t-1])
    return F
```

The `single_exp_smoothing` function set up with the initial forecasted value is assigned as the first value of the series. Let's first evaluate extreme cases of forecasting with $\alpha = 0$ and $\alpha = 1$:

```python
# Single exponential smoothing forecasting
ibm_df['SES0'] = single_exp_smoothing(ibm_df['Close_Price'], 0)
ibm_df['SES0'] = single_exp_smoothing(ibm_df['Close_Price'], 1)
```

The output from the preceding script is as follows:

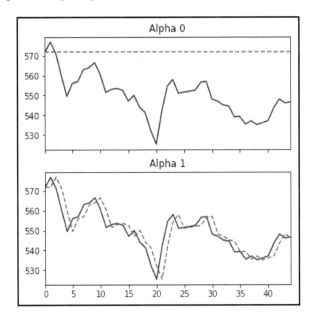

Figure 3.7: Simple exponential smoothing with smoothing parameter of zero and one

The preceding figure illustrates that, at $\alpha = 0$, the forecasted value is a constant, and for $\alpha = 1$, the forecasted series shifts by *1* time lag. The single smoothing forecast for a smoothing value 0.2 can be evaluated as follows:

```
# Single exponential smoothing forecasting
ibm_df['SES'] = single_exp_smoothing(ibm_df['Close_Price'], 0.8)
```

The outcome of the exponential smoothing can be plotted against actual data, as shown in *Figure 3.18*:

```
### Plot Single Exponential Smoothing forecasted value
fig = plt.figure(figsize=(5.5, 5.5))
ax = fig.add_subplot(2,1,1)
ibm_df['Close_Price'].plot(ax=ax)
ax.set_title('IBM Common Stock Close Prices during 1962-1965')
ax = fig.add_subplot(2,1,2)
ibm_df['SES'].plot(ax=ax, color='r')
ax.set_title('Single Exponential Smoothing')
```

Figure 3.8: Comparison between the actual and forecasted values

The accuracy of the model can be evaluated on a hold-out sample using standard objective functions such as **mean square error** (**MSE**) or **mean absolute error** (**MAS**):

$$MSE = \frac{1}{N}\sum_{i=1}^{N}(x_t - F_t)^2$$

Similarly, MAS is evaluated as follows:

$$MAD = \frac{1}{N}\sum_{i-1}^{N}|x_t - F_t|$$

Let's evaluate the effect of α on fitting. To evaluate it, multiple models are developed using different smoothing parameters, as shown here:

```
#Calculate the single exponential forecast at different values
ibm_df['SES2']  = single_exp_smoothing(ibm_df['Close_Price'], 0.2)
ibm_df['SES6']= single_exp_smoothing(ibm_df['Close_Price'], 0.6)
ibm_df['SES8']= single_exp_smoothing(ibm_df['Close_Price'], 0.8)
```

Figure 3.5 compares the forecasting values against the actual ones:

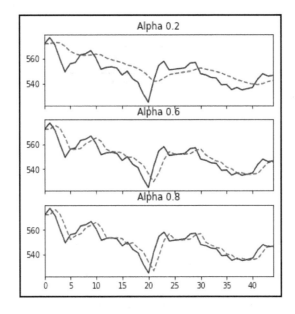

Figure 3.9: Illustration of the effect of *alpha*

The preceding figure illustrates that alpha has a huge impact on forecasting; thus, getting the right alpha value is critical while setting up forecasting. The preceding plot can be obtained using the following script:

```
# Plot the curves
f, axarr = plt.subplots(3, sharex=True)
f.set_size_inches(5.5, 5.5)

ibm_df['Close_Price'].iloc[:45].plot(color='b', linestyle = '-',
```

```
ax=axarr[0])
ibm_df['SES2'].iloc[:45].plot(color='r', linestyle = '-', ax=axarr[0])
axarr[0].set_title('Alpha 0.2')

ibm_df['Close_Price'].iloc[:45].plot(color='b', linestyle = '-',
ax=axarr[1])
ibm_df['SES6'].iloc[:45].plot(color='r', linestyle = '-', ax=axarr[1])
axarr[1].set_title('Alpha 0.6')

ibm_df['Close_Price'].iloc[:45].plot(color='b', linestyle = '-',
ax=axarr[2])
ibm_df['SES8'].iloc[:45].plot(color='r', linestyle = '-', ax=axarr[2])
axarr[2].set_title('Alpha 0.8')
```

As the smoothing helps to reduce the dataset variance, it will reduce the variance of forecasted series between zero to actual variance of dataset:

$$Var(F_T) = \mathrm{var}\left(\alpha \sum_{i=0}^{\infty} (1-\alpha)^i x_{T-i} \right)$$

Solving the previous equation will lead to the following variance:

$$\mathrm{var}(F_T) = \frac{\alpha}{2-\alpha} \mathrm{var}(x_T)$$

Here, T is the length of the time series. For unit variance of series x_T, the variance captured by the forecasted series will vary based on the smoothing parameter α as follows:

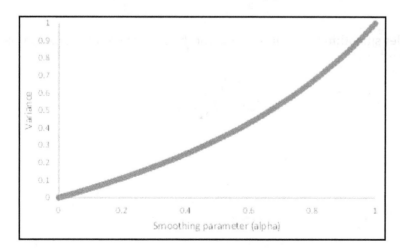

Figure 3.10: Variance captured by simple exponential smoothing with varying *alpha*

Second order exponential smoothing

If first order exponential smoothing does not perform well, then there is a trend in the time series data. The trend is commonly observed in many domains such as when marketing campaigns are run by e-commerce companies, the sales rise or any good annual performance by a company will have a bullish effect on its stock prices. The linear trend can occur due to linear trend between time and response:

$$x_t = constant + \omega t + \varepsilon_t$$

Here, ω is the coefficient that leads to trend. The second order exponential smoothing helps capture the trend in time series data by including another term to the first order exponential smoothing as follows:

$$F_{t+1} = \alpha x_t + (1 - \alpha)(F_{t-1} + T_{t-1})$$

Here, T_t captures the trend component of the exponential smoothing and is represented as follows:

$$T_{t+1} = \beta(F_t - F_{t-1}) + (1 - \beta)(T_{t-1})$$

Here, α is the data smoothing factor and β is the trend smoothing factor with values between [0, 1]. The next stage forecast can be generated as follows:

$$\hat{x}_{t+1} = F_t + T_t$$

In second order smoothing, the initial value for the trend component can be assigned in multiple ways:

$$T_t = x_2 - x_1$$
$$T_t = (x_n - x_1)/(n - 1)$$

Here, *n* is the number of observations. Let's demonstrate an example of second order smoothing. We will use beer production data. The following are the steps of demonstration:

1. The first step is to load the required modules:

```
# Load modules
from __future__ import print_function
import os
import pandas as pd
import numpy as np
from matplotlib import pyplot as plt

Load beer dataset as pandas DataFrame.
#Read dataset into a pandas.DataFrame
beer_df = pd.read_csv('datasets/quarterly-beer-production-in-
aus-March 1956-June 1994.csv')
```

2. Let's create a function for double exponential smoothing:

```
# Function for double exponential smoothing
def double_exp_smoothing(x, alpha, beta):
    yhat = [x[0]] # first value is same as series
    for t in range(1, len(x)):
        if t==1:
            F, T= x[0], x[1] - x[0]
        F_n_1, F = F, alpha*x[t] + (1-alpha)*(F+T)
        T=beta*(F-F_n_1)+(1-beta)*T
        yhat.append(F+T)
    return yhat
```

> The preceding function takes time series as an input with alpha and beta. The preceding implementation uses the difference of the first two occurrences to set up the initial trend value.

3. Let's evaluate the performance on boundary cases of alpha beta, that is, (0,0), (0, 1), (1,0), and (1,1) values of alpha and beta smoothing parameters:

```
# Effect of alpha and beta
beer_df['DEF00'] = double_exp_smoothing(beer_df['Beer_Prod'],0, 0)
beer_df['DEF01'] = double_exp_smoothing(beer_df['Beer_Prod'],0, 1)
beer_df['DEF10'] = double_exp_smoothing(beer_df['Beer_Prod'],1, 0)
beer_df['DEF11'] = double_exp_smoothing(beer_df['Beer_Prod'],1, 1)
```

The outcome from the preceding script to fit second order exponential smoothing is shown in the following figure:

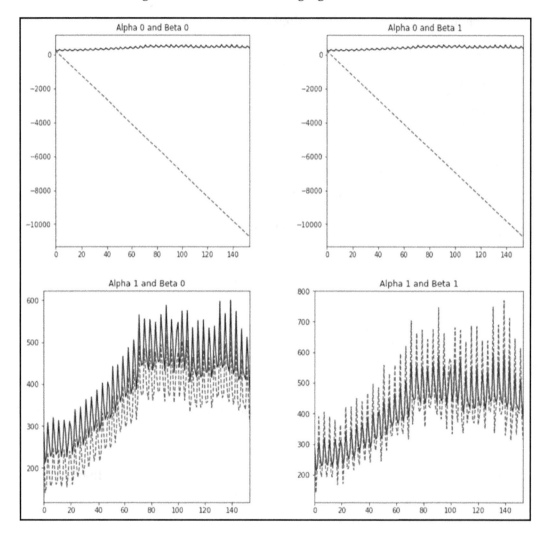

Figure 3.11: Effect of alpha and beta smoothing parameters in second order exponential smoothing on the beer sales dataset

When $\alpha = 0$, the initial values remain constant; thus, the trend parameter did not play a role. However, as $\alpha = 1$ and $\beta = 1$, the second order exponential smoothing can be written as follows:

$$F_t = x_{t-1}$$
$$T_t = T_{t-1}$$
$$\hat{x}_t = x_{t-1} + T_{t-1}$$

The forecast at time t depends on the previous value and trend components. As β is set to zero, the trend component at t-1 will depend on t-2:

$$T_t = T_{t-1} = T_{t-2} = T_{t-3} = \ldots = T_0$$

Thus, the trend component value depends on the initial assigned value and is a constant.

Similarly, for $\alpha = 1$ and $\beta = 1$, the second order exponential smoothing can be simplified as follows:

$$F_t = x_{t-1}$$
$$T_t = (F_{t-1} - F_{t-2}) = x_{t-2} - x_{t-3}$$
$$\hat{x}_t = x_{t-1} + (x_{t-2} - x_{t-3})$$

With $\alpha = 1$ and $\beta = 1$, the difference of t-2 and t-3 is added to time t forecast value as compared to the configuration of $\alpha = 1$ and $\beta = 0$, which gives the shift to the prediction and is closer to the real forecast.

Let's perform double exponential smoothing on the beer data using intermediate values of α and β as follows:

```
beer_df['DEF'] = double_exp_smoothing(beer_df['Beer_Prod'], 0.4, 0.7)
```

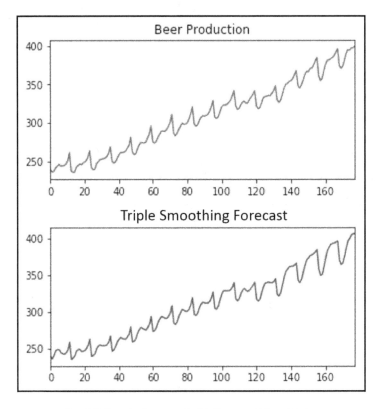

Figure 3.12: Comparison between actual and double exponential forecasted values

Let's also compare the performance of single and double exponential smoothing:

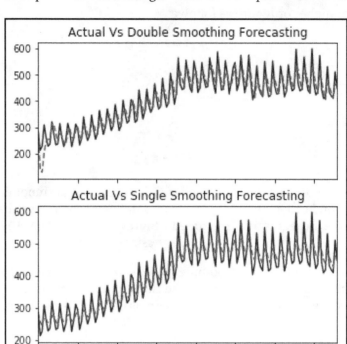

Figure 3.13: Comparison between single and double exponential forecasted values

The preceding figure shows that double exponential smoothing is able to capture the variation of the real signal better for the current dataset as compared to single exponential smoothing. However, in scenarios where the trend component tends to zero, the performance of the single and double exponential smoothing approach is comparable.

Modeling higher-order exponential smoothing

The concept can be further extended to higher-order exponential smoothing with an nth order polynomial model:

$$x_t = \alpha_0 + \alpha_1 t + \frac{\alpha_2}{2!} t^2 + \ldots + \frac{\alpha_n}{n!} t^n + \varepsilon_t$$

Here, error $\varepsilon_t \sim N(0,\sigma^2)$ is normally distributed with 0 mean and σ variance. The exponential smoothers used for higher order are as follows:

$$\tilde{x}_t^{(1)} = \kappa x_t^{(1)} + (1-\kappa)\tilde{x}_{t-1}^{(1)}\alpha_1 t$$

$$\tilde{x}_t^{(2)} = \kappa x_t^{(2)} + (1-\kappa)\tilde{x}_{t-1}^{(2)}\alpha_1 t$$

$$\ldots$$

$$\tilde{x}_t^{(n)} = \kappa x_t^{(n-1)} + (1-\kappa)\tilde{x}_{t-1}^{(n)}$$

Here, is weights for smoothers. Usually, higher-order exponential smoothing is not used in time as even for second order smoothing, the computation becomes very hard and approaches such as **Autoregressive Integrated Moving Average (ARIMA)** are utilized. It will be further discussed in Chapter 4, *Auto Regressive Models*.

Another very popular exponential smoothing is triple exponential smoothing. The triple exponential smoothing allows you to capture seasonality with level and trend. The relationship between levels, trends, and seasonality is defined using the following set of equations:

$$F_t = \alpha(x_t - S_{t-L}) + (1-\alpha)(F_{t-1} + T_{t-1})$$

In these equations, F_t captures levels of observation at time t. Similarly, T_t and S_t captures trend and seasonality at time t. The coefficients α, β and γ represent the data smoothing factor, trend smoothing factor, and seasonality smoothing factor, respectively, with values between [0,1]. These equations can be used to forecast the next time period as follows:

$$T_t = \beta(F_t - F_{t-1}) + (1-\beta)(T_{t-1})$$

The preceding equation can be generalized for any period m as follows:

$$S_t = \gamma(x_t - F_t) + (1-\gamma)S_{t-c}$$

The term F_t captures the offset of seasonal components from the last observed seasonal trend. As triple exponential smoothing is applied for seasonal and trend time series, the values can be utilized to compute better starting values for the trend. Let's demonstrate triple order exponential smoothing using Wisconsin Employment data.

The Wisconsin Employment data can be downloaded from `http://datamarket.com` and the csv file can be loaded in the Python environment using the `pandas.read_csv` command.

1. The first step is to load the required modules:

```
# Load modules
from __future__ import print_function
import os
import pandas as pd
import numpy as np
from matplotlib import pyplot as plt

#read the data from into a pandas.DataFrame
wisc_emp = pd.read_csv('datasets/wisconsin-employment-time-series.csv')
```

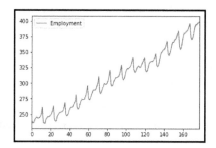

Figure 3.14: Wisconsin employee dataset

The preceding figure shows Wisconsin employee time series dataset. The dataset consists of both annual trend and monthly seasonality. We are aware about the seasonality pattern for data. Thus, seasonality information can be used to derive the initial value of trend as the average value across seasons using the following equation:

$$T_0 = \frac{1}{C}(\frac{x_{C+1} - x_1}{C} + \frac{x_{C+2} - x_2}{C} + \frac{x_{C+3} - x_3}{C} \dots)$$

1. The preceding equation can be implemented in Python as follows:

```
# Initialize trend value
def initialize_T(x, seasonLength):
    total=0.0
    for i in range(seasonLength):
        total+=float(x[i+seasonLength]-x[i])/seasonLength
    return total
```

For example, the initial trend value generated by the preceding function is 1.69 using the following script:

```
initialize_T(wisc_emp['Employment'], 12)
```

2. The initial seasonality is very critical and can be calculated using the following function:

```
# Initialize seasonal trend
def initialize_seasonalilty(x, seasonLength):
    seasons={}
    seasonsMean=[]
    num_season=int(len(x)/seasonLength)
    # Compute season average
    for i in range(num_season):
seasonsMean.append(sum(x[seasonLength*i:seasonLength*i+seasonLength])/float
(seasonLength))
    # compute season intial values
    for i in range(seasonLength):
        tot=0.0
        for j in range(num_season):
            tot+=x[seasonLength*j+i]-seasonsMean[j]
        seasons[i]=tot/num_season
    return seasons
```

The initial values of seasons are calculated as mean value of response x.

3. Once values are obtained, we are ready to set up triple exponential forecasting:

```
# Triple Exponential Smoothing Forecast
def triple_exp_smoothing(x, seasonLength, alpha, beta, gamma, h):
    yhat=[]
    S = initialize_seasonalilty(x, seasonLength)
    for i in range(len(x)+h):
        if i == 0:
            F = x[0]
            T = initialize_T(x, seasonLength)
            yhat.append(x[0])
            continue
        if i >= len(x):
            m = i - len(x) + 1
            yhat.append((F + m*T) + S[i%seasonLength])
        else:
            obsval = x[i]
            F_last, F= F, alpha*(obsval-S[i%seasonLength]) + (1-
alpha)*(F+T)
            T = beta * (F-F_last) + (1-beta)*T
            S[i%seasonLength] = gamma*(obsval-F) + (1-
```

```
gamma)*S[i%seasonLength]
            yhat.append(F+T+S[i%seasonLength])
    return yhat
```

The triple exponential smoothing is controlled by α, β and γ. The presence or absence of any scenario will have a drastic effect on the outcome. Let's have an empirical comparison for different extreme scenarios, as shown in the Wisconsin employment dataset:

S. No.	Alpha	Beta	Gamma
1	0	0	1
2	0	1	0
3	1	0	0
4	1	1	0
5	1	0	1
6	0	1	1
7	1	1	1

Figure 3.15: Different configurations at extreme values

The forecast for triple exponential smoothing using either alpha, beta, or gamma values set to one is generated using the following code:

```
# Effect of alpha and beta
wisc_emp['TEF001'] = triple_exp_smoothing(wisc_emp['Employment'], 12, 0, 0,
1, 0)
wisc_emp['TEF010'] = triple_exp_smoothing(wisc_emp['Employment'], 12, 0, 1,
0, 0)
wisc_emp['TEF100'] = triple_exp_smoothing(wisc_emp['Employment'], 12, 1, 0,
0, 0)

# Plot alpha=0, beta=0, gamma=1
fig = plt.figure(figsize=(5.5, 5.5))
ax = fig.add_subplot(3,1,1)
wisc_emp['Employment'].plot(color='b', linestyle = '-', ax=ax)
wisc_emp['TEF001'].plot(color='r', linestyle = '--', ax=ax)
ax.set_title('TES: alpha=0, beta=0, gamma=1')

# Plot alpha=0, beta=1, gamma=0
ax = fig.add_subplot(3,1,2)
wisc_emp['Employment'].plot(color='b', linestyle = '-', ax=ax)
wisc_emp['TEF010'].plot(color='r', linestyle = '--', ax=ax)
ax.set_title('TES: alpha=0, beta=1, gamma=0')

# Plot alpha=1, beta=0, gamma=0
```

```
ax = fig.add_subplot(3,1,3)
wisc_emp['Employment'].plot(color='b', linestyle = '-', ax=ax)
wisc_emp['TEF100'].plot(color='r', linestyle = '--', ax=ax)
ax.set_title('TES: alpha=1, beta=0, gamma=0')
fig.subplots_adjust(hspace=.5)
```

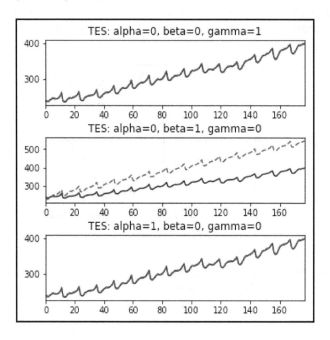

Figure 3.16: Forecasting with triple exponential smoothing with (i) alpha:0, beta:0, and gamma:1; (ii) alpha:0, beta:1, and gamma:0; and (iii) alpha:0, beta:0, and gamma:1

The preceding figure shows triple exponential smoothing using either data smoothing factor alpha or seasonality factor gamma set to one. Similarity plotting with respect to two parameters set to one is shown in the following figure using this script:

```
# Effect of alpha and beta
wisc_emp['TEF110'] = triple_exp_smoothing(wisc_emp['Employment'], 12, 1, 1,
0, 0)
wisc_emp['TEF101'] = triple_exp_smoothing(wisc_emp['Employment'], 12, 1, 0,
1, 0)
wisc_emp['TEF011'] = triple_exp_smoothing(wisc_emp['Employment'], 12, 1, 1,
1, 0)
```

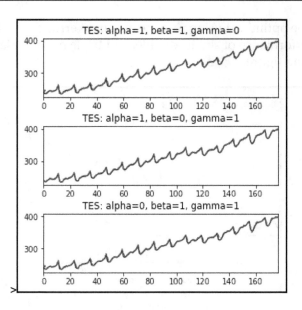

Figure 3.17: Forecasting with triple exponential smoothing with (i) alpha:1, beta:1, and gamma:0; (ii) alpha:1, beta:0, and gamma:1; and (iii) alpha:0, beta:1, and gamma:1

The model without beta set to zero outperforms other models. Let's run triple exponential smoothing with intermediate parameters, as shown in the following figure:

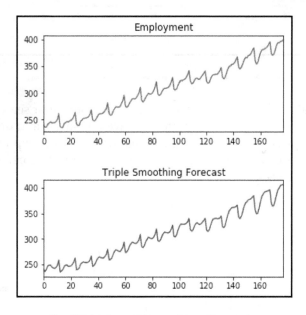

Figure 3.18: Actual versus triple exponential smoothing comparison

The forecasting can be optimized using root mean square error on the hold-out sample. Let's compare the performance of single, double, and triple exponential smoothing, as shown in the following figure:

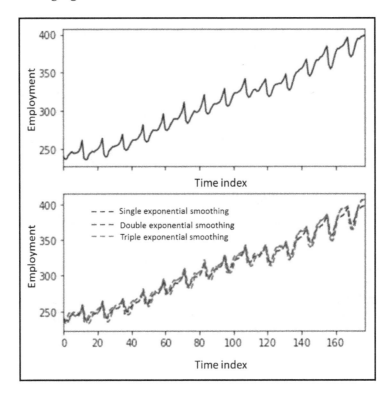

Figure 3.19: Comparison between single, double, and triple exponential smoothing

As based on empirical studies, the single level using smoothing or seasonality is able to capture the data trend; thus, all models performed well as single and double exponential were able to use the smoothing factor to do forecasting and triple exponential smoothing is able to capture forecasts using either smoothing or seasonality factors.

Summary

This chapter covers exponential smoothing approaches to smoothen time series data. The approaches can be easily extended for the forecasting by including terms such as smoothing factor, trend factor, and seasonality factor. The single order exponential smoothing performs smoothing using only the smoothing factor, which is further extended by second order smoothing factor by including the trend component. The third order smoothing was also covered, which incorporates all smoothing, trend, and seasonality factors into the model.

This chapter covered all these models in detail with their Python implementation. The smoothing approaches can be used to forecast if the time series is a stationary signal. However, the assumption may not be true. Higher-order exponential smoothing is recommended but its computation becomes hard. Thus, to deal with the approach, other forecasting techniques such as ARIMA is proposed, which will be covered in the next chapter.

4
Auto-Regressive Models

In the previous chapter, exponential smoothing-based forecasting techniques were covered, which is based on the assumption that time series is composed on deterministic and stochastic terms. The random component is zero out with number of observations considered for the forecasting. This assumes that random noise is truly random and follows independent identical distribution. However, this assumption often tends to get violated and smoothing is not sufficient to model the process and set up a forecasting model.

In these scenarios, auto-regressive models can be very useful as these models adjust immediately using the prior lag values by taking advantage of inherent serial correlation between observations. This chapter introduces forecasting concepts using auto-regressive models. The auto-regressive model includes auto-regressive terms or moving average terms. Based on the components used, there are multiple approaches that can be used in time series forecasting such as **moving average (MA)**, **auto-regressive moving average (ARMA)**, and **auto-regressive integrated moving average (ARIMA)**. The MA in this chapter is different from the moving average smoothing discussed in Chapter 2, *Understanding Time Series Data*. The MA, or more appropriately MA(q) with order q, is an auto-regressive moving average model based on error lag regression.

The current chapter focuses on auto-regresive model and will cover following topics:

- Moving Average (MA)
- Auto-regressive (AR)
- Auto-regressive moving average (ARMA)
- Auto-regressive integrated moving average (ARIMA)
- Summary
- Introduction

The concept of auto-regressive models in time series is referred to models that are developed by regressing on previous values. For example, x_t is response at time t and the model is developed as follows:

$$x_t = \emptyset \, x_{t-1} + \mathcal{E}$$

The preceding equation is a simple example of an AR(1) model. Here, $\emptyset$ is the model coefficient and $\mathcal{E}$ is the error. Additionally, similar to the regression model, error normality assumption stays for an auto-regressive model as well as consideration of stationarity or homoscedastic. The next subsection introduces the moving average models, which is a linear dependence on historical deviation of models from the last prior value.

Auto-regressive models

Another very famous approach to regress on time series data is to regress it with its lag term. This genre of models is referred to as **auto-regressive models** (**AR models**). The AR models are very good in capturing trends as the next time values are predicted based on the prior time values. Thus, AR models are very useful in situations where the next forecasted value is a function of the previous time period, such as an increase in average stock price gain due to good company growth; we expect the effect to retain over time and price should keep increasing as a function of time as the trend component.

The auto-regressive model is defined as AR(p), where p refers to the order of the AR component.

The first-order AR model is denoted by AR(1):

$$X_t = \phi \epsilon_{t-1} + \epsilon_t$$

The second-order AR model is denoted by AR(2):

$$X_t = \phi_1 \epsilon_{t-1} + \phi_2 \epsilon_{t-2} + \epsilon_t$$

The pth order AR model is denoted by AR(p):

$$X_t = \phi_1 \epsilon_{t-1} + \phi_2 \epsilon_{t-2} + \ldots + \phi_p \epsilon_{t-p} + \epsilon_t$$

Here, ϕ is the model coefficient, $\epsilon_t \sim N(0, \sigma^2)$ is an error in time t, and p is the order of the AR model. Let's utilize a setup similar to the one used for moving average models to understand the modeling implications of AR components. The AR(1) dataset can be generated using the `arma_generate_sample` function from the *statsmodels.tsa* module:

```
import statsmodels.tsa.api as smtsa
# Number of samples
n = 600
# Generate AR(1) dataset
ar = np.r_[1, -0.6]
ma = np.r_[1, 0]
ar1_data = smtsa.arma_generate_sample(ar=ar, ma=ma, nsample=n)
plotds(ar1_data)
```

The preceding script generates a dataset for the AR(1) scenario with serial correlation defined for the previous lag as 0.6. The MA component is set to zero to remove any moving average effect from the time series signal. The time series signal generated and autocorrelation and partial autocorrelation for the generated signal is shown in Figure 4.1:

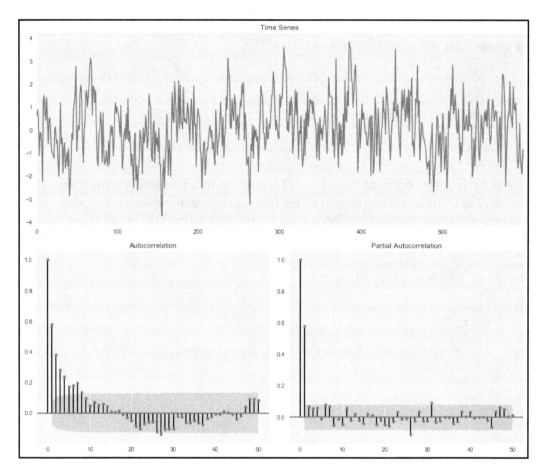

Figure 4.1: AR(1) time series signal with its ACF and PACF plots

The model relationship in Figure 4.7 can be represented as $x_t = \phi_1 x_{t-1} + \epsilon_t$ as the data is simulated using the *AR* component of less than 1; thus, auto correlation will decrease over time following this relationship between ϕ and t:

$$X_t = \phi_1 X_{t-1} + \epsilon_{t=} \phi_1(\phi_1 X_{t-1} + \epsilon_{t-1}) + \epsilon_t = \phi_1^2 X_{t-2} + \phi_1 \epsilon_{t-1} + \epsilon_t$$

Thus, the ACF graph decreases exponentially whereas as PACF removes the lag effect while computing correlation, only significant terms are captured. The Φ value affects the signal stationarity. For example, if we increase Φ from 0.6 to 0.95 in AR(1), the model tends toward non-stationarity, as shown in the following image:

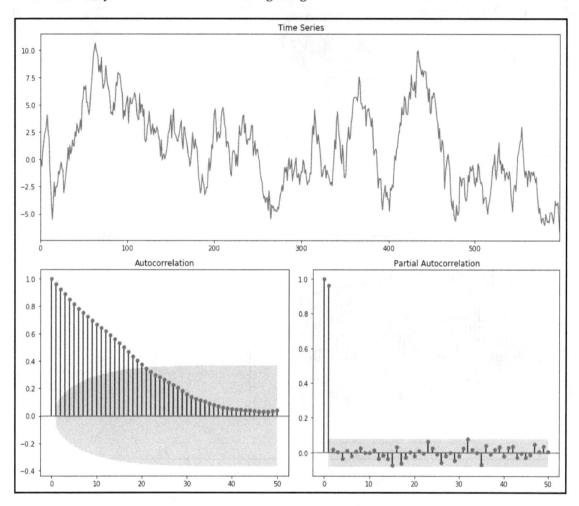

Figure 4.2: AR(1) time series signal with high autocorrelation of 0.95

In scenario Φ>1, the model becomes non-stationary. An example of non-stationary process with Φ>1 is shown here:

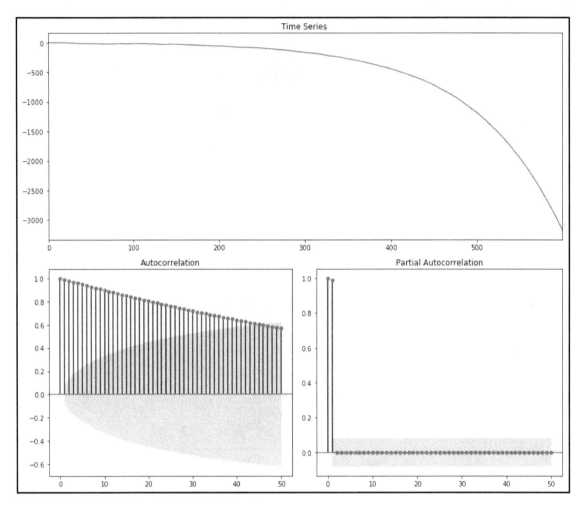

Figure 4.3: AR(1) time series signal with >1 leading to non-stationary signal

Figure 4.9 is generated with Φ = 1.01. Similarly, higher-order AR models can be generated to validate effects on the PACF component with order. The datasets with AR(2) and AR(3) components are generated using the following script:

```
# Generate AR(2) dataset
ar = np.r_[1, 0.6, 0.7]
ma = np.r_[1, 0]
ar2_data = smtsa.arma_generate_sample(ar=ar, ma=ma, nsample=n)
```

```
plotds(ar2_data)

# Generate AR(3) dataset
ar = np.r_[1, 0.6, 0.7, 0.5]
ma = np.r_[1, 0]
ar3_data = smtsa.arma_generate_sample(ar=ar, ma=ma, nsample=n)
plotds(ar3_data)
```

The time series signal with their ACF and PACF plots for AR(2) and AR(3) time series signals are shown in Figures 4.4 and 4.5:

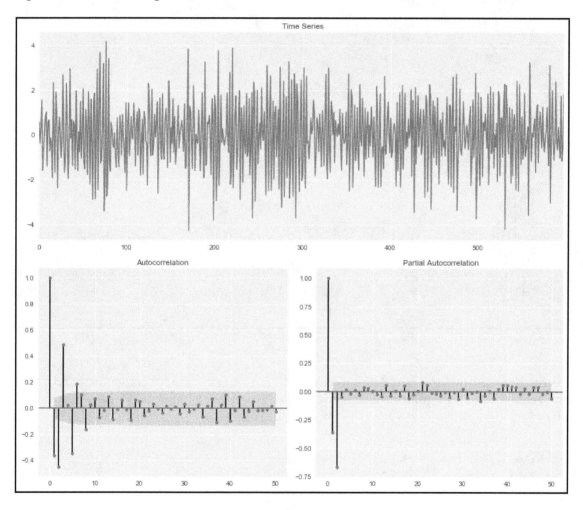

Figure 4.4: AR(2) time series with its ACF and PACF

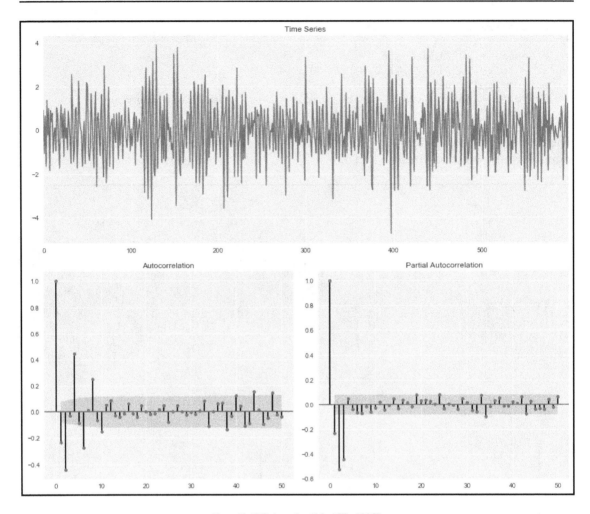

Figure 4.5: AR(3) time series with its ACF and PACF

As can be seen from different data sources, PACF is capturing the AR component and q is the value where it is significant. The model for AR1 can be evaluated using the ARMA.fit class from the statsmodels.tsa.api module in Python;

```
# Build AR(1) model
ar1model = smtsa.ARMA(ar1_data.tolist(), order=(1, 0))
ar1=ar1model.fit(maxlag=30, method='mle', trend='nc')
ar1.summary()
```

The preceding script fits an AR(1) model on the time series dataset and the outcome is reported as follows:

```
In [167]: ar1 = smtsa.ARMA(ar1_data.tolist(), order=(1, 0)).fit(
     ...:      maxlag=30, method='mle', trend='nc')
     ...: ar1.summary()
Out[167]:
<class 'statsmodels.iolib.summary.Summary'>
"""
                         ARMA Model Results
==============================================================================
Dep. Variable:                      y   No. Observations:              600
Model:                     ARMA(1, 0)   Log Likelihood             -849.954
Method:                           mle   S.D. of innovations           0.997
Date:                Thu, 17 Aug 2017   AIC                        1703.908
Time:                        07:46:56   BIC                        1712.702
Sample:                             0   HQIC                       1707.331

==============================================================================
                 coef    std err          z      P>|z|      [0.025      0.975]
------------------------------------------------------------------------------
ar.L1.y        0.5802      0.033     17.468      0.000       0.515       0.645
                                  Roots
==============================================================================
                  Real          Imaginary           Modulus         Frequency
------------------------------------------------------------------------------
AR.1            1.7235           +0.0000j            1.7235            0.0000
------------------------------------------------------------------------------
"""
```

Figure 4.6: AR(1) model output

The AR(1) is simulated using lag serial correlation of 0.6 and the fitted value is evaluated as 0.58, which is quite close to the actual relationship. Similarly, the AR(3) model is fitted on a generated dataset with AR with an actual generated lag relationship of 0.6, 0.7, and 0.5 and the following is the outcome of the fitted value:

```
Out[168]:
<class 'statsmodels.iolib.summary.Summary'>
"""
                          ARMA Model Results
==============================================================================
Dep. Variable:                     y   No. Observations:                  600
Model:                     ARMA(3, 0)   Log Likelihood                -852.586
Method:                          mle   S.D. of innovations              1.001
Date:                Thu, 17 Aug 2017   AIC                           1713.172
Time:                       07:48:10   BIC                           1730.759
Sample:                            0   HQIC                          1720.018

=================================================================================
                 coef    std err          z      P>|z|      [0.025      0.975]
---------------------------------------------------------------------------------
ar.L1.y       -0.5872      0.037    -16.045      0.000      -0.659      -0.516
ar.L2.y       -0.6788      0.034    -20.189      0.000      -0.745      -0.613
ar.L3.y       -0.4446      0.037    -12.161      0.000      -0.516      -0.373
                                   Roots
=================================================================================
                  Real          Imaginary           Modulus         Frequency
---------------------------------------------------------------------------------
AR.1            0.0303           -1.1900j            1.1904           -0.2460
AR.2            0.0303           +1.1900j            1.1904            0.2460
AR.3           -1.5874           -0.0000j            1.5874           -0.5000
---------------------------------------------------------------------------------
"""
```

Figure 4.7: AR(3) model output

The AR model fitted shows a relationship of 0.58, 0.67 ,and 0.44, which is quite close to the real relationship. Both AR and MA can be used to correct the serial dependency but usually, positive autocorrelation is corrected using AR models and negative dependency is corrected using MA models.

Moving average models

The moving average models use dependency between residual errors to forecast values in the next time period. The model helps you adjust for any unpredictable events such as catastrophic events leading to a share market crash leading to share prices falling, which will happen over time and is captured as a moving average process.

The first-order moving average denoted by MA(1) is as follows:

$$x_t = \alpha - \theta_1 \epsilon_{t-1} + \epsilon_t$$

The second-order moving average denoted by MA(2) is as follows:

$$x_t = \alpha - \theta_1 \epsilon_{t-1} - \theta_2 \epsilon_{t-2} + \epsilon_t$$

The qth order moving average denoted by MA(q) is as follows:

$$x_t = \alpha - \theta_1 \epsilon_{t-1} - \theta_2 \epsilon_{t-2} - ... - \theta_q \epsilon_{t-q} + \epsilon_t$$

Here, ϵ_t is the identically independently-distributed error at time t and follows normal distribution $N(0, \sigma^2_\epsilon)$ with zero mean and σ^2_ϵ variance. The ϵ_t component represents error in time t and the α and ϵ notations represent mean intercept and error coefficients, respectively. The moving average time series model with q^{th} order is represented as MA(q). The preceding relations do not change the expectation value for MA(q), which is defined as follows:

$$E(x_t) = E(\alpha - \theta_1 \epsilon_{t-1} - \theta_2 \epsilon_{t-2} - ... - \theta_q \epsilon_{t-q} + \epsilon_t) = \alpha$$

However, the variance is increased and is defined as follows:

$$var(x_t) = var(\alpha - \theta_1 \epsilon_{t-1} - \theta_2 \epsilon_{t-2} - ... - \theta_q \epsilon_{t-q} + \epsilon_t) = \alpha$$

$$var(x_t) = \sigma^2 (1 + \theta_{1^2} + \theta_{2^2} + ... \theta_{q^2})$$

To illustrate a moving average time series model, let's generate a signal using the following snippet:

```
import statsmodels.tsa.api as smtsa
# Number of samples
n = 600
# Generate MA(1) dataset
ar = np.r_[1, -0]
ma = np.r_[1, 0.7]
ma1_data = smtsa.arma_generate_sample(ar=ar, ma=ma, nsample=n)
```

In the preceding snippet, *n* represents the number of samples to be generated with *ar* defining the auto-regressive component and *ma* explains the moving average component of the time series signal. The details on the *ar* component will be covered in the auto-regressive section of `Chapter 4`, *Auto Regressive Models*. Currently, we will keep the impact of *ar* on the time series signal as zero. The preceding snippet will generate a time series dataset with an MA(1) dependency with 0.7 serial correlation between the error and can be represented as follows:

$$x_t = 0.7\epsilon_{t-1} + \epsilon_t$$

The preceding script will generate the following signal:

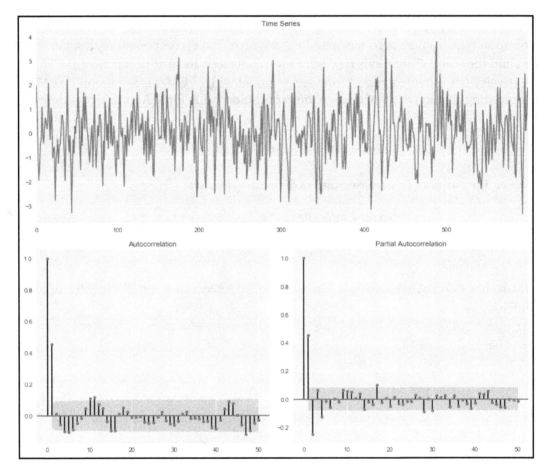

Figure 4.8: Example of MA(1) dataset

To evaluate whether the time series signal consists of an MA or AR component, autocorrelation (ACF) and partial autocorrelation (PACF) is used. The ACF represents:

$$ACF = \frac{\text{cov}(x_t, x_{t-h})}{\text{var}(x_t)}$$

The numerator in the preceding equation is covariance between the time series signal at time t and t-h where h is the lag in the time series signal. The PACF is also computed similarly as ACF except that correlation is computed by removing the already explained variation between intervals. This is also defined as conditional correlation. The first-order ACF is similar to first-order PACF. In second-order (lag) PACF, the condition probability starts playing a significant role:

$$PACF = \frac{\text{cov}(x_t, x_{t-2}|x_{t-1})}{\sqrt{\text{var}(x_t|x_{t-1})}\sqrt{\text{var}(x_{t-2}|x_{t-1})}}$$

Similarly, third-order PACF can be represented as follows:

$$PACF = \frac{\text{cov}(x_t, x_{t-3}|x_{t-1}, x_{t-2})}{\sqrt{\text{var}(x_t|x_{t-1}, x_{t-2})}\sqrt{\text{var}(x_{t-3}|x_{t-1}, x_{t-2})}}$$

The preceding relation can be further extended to higher-order lags. Let's have a look at ACF and PACF functions for the previously generated time series:

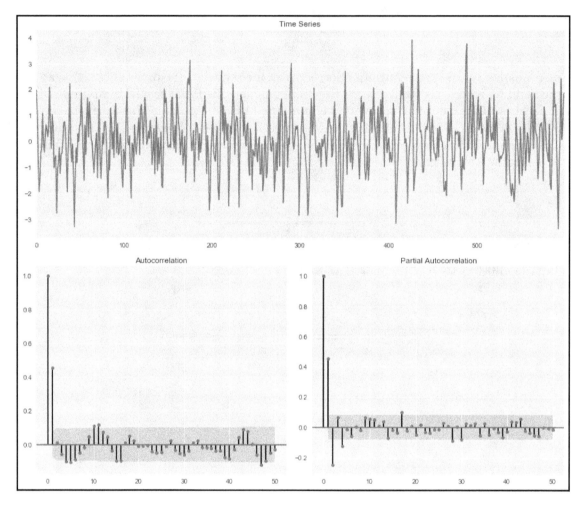

Figure 4.9: ACF and PACF for MA(1) dataset

The ACF on the preceding dataset shows a 1-lag dependency. As the MA relationship is captured using $x_t = \theta \epsilon_{t-1} + \epsilon_t$, which is independent of lag terms, the ACF tends to capture the appropriate order q of the MA series. As can be seen from Figure 4.2, the ACF does not go to zero after the defined order, rather it reduces to a small value. The confidence interval is tested using relation $\frac{\pm 2}{\sqrt{N}}$, where N and $\frac{1}{\sqrt{N}}$ represent an approximation of the standard deviation ,which is true under the independence condition.

Let's see the impact of the MA component of ACF and PACF with a higher order of q using the following script:

```
# Generate MA(2) dataset
ar = np.r_[1, -0]
ma = np.r_[1, 0.6, 0.7]
ma2_data = smtsa.arma_generate_sample(ar=ar, ma=ma, nsample=n)
plotds(ma2_data)

# Generate MA(3) dataset
ar = np.r_[1, -0]
ma = np.r_[1, 0.6, 0.7, 0.5]
ma3_data = smtsa.arma_generate_sample(ar=ar, ma=ma, nsample=n)
plotds(ma3_data)
```

The preceding script will generate a moving average time series signal of order MA(2) and MA(3) having no impact of the AR component. The ACF and PACF plots for time series signals generated using the preceding scripts are shown here:

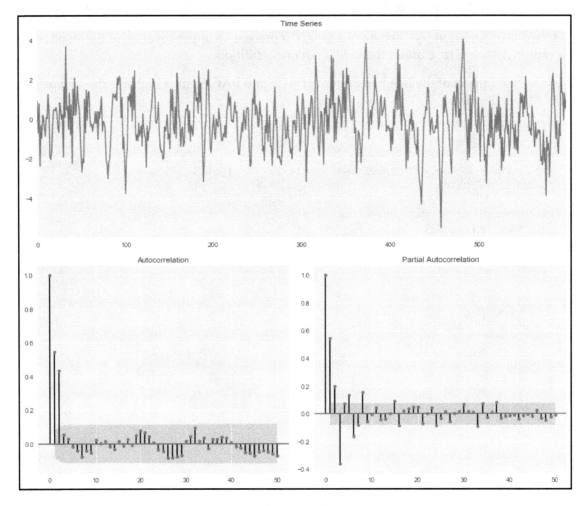

Figure 4.10: Example of MA(2) dataset

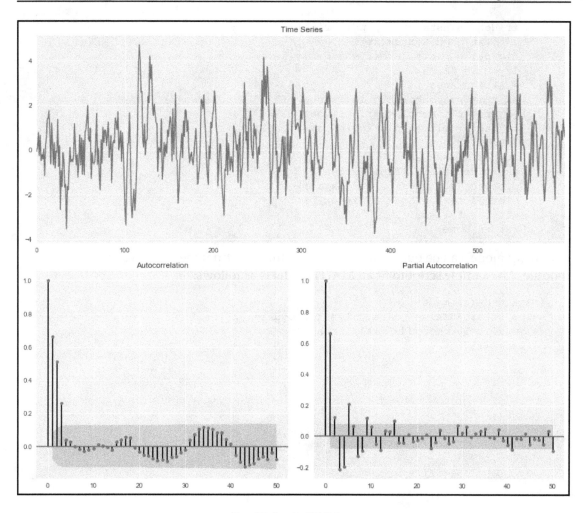

Figure 4.11: Example of MA(3) dataset

Usually, ACF defines error serial correlation well and is thus used to detect MA(q); however, as the order increases and other time series components such as seasonality, trend, or stationary come into picture, this makes interpretation harder. The MA(q) assumes that the process is stationary and error is white noise to ensure unbiased estimation. For example, in the preceding MA(2) and MA(3), the seasonality component in the data makes the interpretation slightly harder. The preceding plots are generated using the following functions:

```
from statsmodels.graphics.tsaplots import plot_acf, plot_pacf
from matplotlib import pyplot as plt
def plotds(xt, nlag=30, fig_size=(12, 10)):
```

```
if not isinstance(xt, pd.Series):
    xt = pd.Series(xt)
fig_plt = plt.figure(figsize=fig_size)
layout = (2, 2)
# Assign axes
ax_xt = plt.subplot2grid(layout, (0, 0), colspan=2)
ax_acf= plt.subplot2grid(layout, (1, 0))
ax_pacf = plt.subplot2grid(layout, (1, 1))
# Plot graphs
xt.plot(ax=ax_xt)
ax_xt.set_title('Time Series')
plot_acf(xt, lags=50, ax=ax_acf)
plot_pacf(xt, lags=50, ax=ax_pacf)
plt.tight_layout()
return None
```

An MA(q) model can be built using the ARMA function from the `statsmodel.tsa` module. An example script to fit an MA(1) model is as follows:

```
# Build MA(1) model
ma1 = smtsa.ARMA(ma1_data.tolist(), order=(0, 1)).fit(
    maxlag=30, method='mle', trend='nc')
ma1.summary()
```

As the order of AR is kept zero, smtsa.ARMA builds an MA(1). The model summary returned by smtsa.ARMA is shown here:

```
In [161]: ma1 = smtsa.ARMA(ma1_data.tolist(), order=(0, 1)).fit(
     ...:        maxlag=30, method='mle', trend='nc')
     ...: ma1.summary()
Out[161]:
<class 'statsmodels.iolib.summary.Summary'>
"""
                          ARMA Model Results
==============================================================================
Dep. Variable:                    y   No. Observations:              600
Model:                   ARMA(0, 1)   Log Likelihood             -833.563
Method:                         mle   S.D. of innovations          0.970
Date:              Wed, 16 Aug 2017   AIC                       1671.125
Time:                      08:46:24   BIC                       1679.919
Sample:                           0   HQIC                      1674.548

==============================================================================
                 coef    std err          z      P>|z|      [0.025      0.975]
------------------------------------------------------------------------------
ma.L1.y        0.6717      0.029     23.172      0.000       0.615       0.728
                                  Roots
==============================================================================
                  Real          Imaginary           Modulus         Frequency
------------------------------------------------------------------------------
MA.1           -1.4889           +0.0000j            1.4889            0.5000
------------------------------------------------------------------------------
"""
```

Figure 4.12: Output from an MA(1) model

As you can see, the model has captured the *0.67* correlation between residuals, which is quite close to the simulated value of *0.7*. Similarly, we run the model for the MA(3) dataset and the outcome is shown in Figure 4.13:

```
In [162]: ma3 = smtsa.ARMA(ma3_data.tolist(), order=(0, 3)).fit(
     ...:      maxlag=30, method='mle', trend='nc')
     ...: ma3.summary()
Out[162]:
<class 'statsmodels.iolib.summary.Summary'>
"""
                            ARMA Model Results
==============================================================================
Dep. Variable:                     y   No. Observations:                  600
Model:                    ARMA(0, 3)   Log Likelihood                -843.388
Method:                          mle   S.D. of innovations              0.985
Date:               Wed, 16 Aug 2017   AIC                           1694.775
Time:                       08:49:09   BIC                           1712.363
Sample:                            0   HQIC                          1701.622

==============================================================================
                 coef    std err          z      P>|z|      [0.025      0.975]
------------------------------------------------------------------------------
ma.L1.y        0.5817      0.036     16.179      0.000       0.511       0.652
ma.L2.y        0.7177      0.031     23.455      0.000       0.658       0.778
ma.L3.y        0.4761      0.037     12.914      0.000       0.404       0.548
                                    Roots
==============================================================================
                  Real          Imaginary           Modulus         Frequency
------------------------------------------------------------------------------
MA.1           0.0349           -1.1535j            1.1540           -0.2452
MA.2           0.0349           +1.1535j            1.1540            0.2452
MA.3          -1.5771           -0.0000j            1.5771           -0.5000
------------------------------------------------------------------------------
```

Figure 4.13: Output from the MA(3) model on a simulated dataset

Building datasets with ARMA

The preceding two sections describe the auto-regressive model AR(p), which regresses on its own lagged terms and moving average model MA(q) builds a function of error terms of the past. The AR(p) models tend to capture the mean reversion effect whereas MA(q) models tend to capture the shock effect in error ,which are not normal or unpredicted events. Thus, the ARMA model combines the power of AR and MA components together. An ARMA(p, q) time series forecasting model incorporates the pth order AR and q^{th} order MA model, respectively.

The ARMA (1, 1) model is represented as follows:

$$x_t = \alpha + \phi_1 x_{t-1} - \theta_1 \varepsilon_{t-1} + \varepsilon_t$$

The ARMA(1, 2) model is denoted as follows:

$$x_t = \alpha + \phi_1 x_{t-1} - \theta_1 \varepsilon_{t-1} - \theta_2 \varepsilon_{t-2} + \varepsilon_t$$

The ARMA(p, q) model is denoted as follows:

$$x_t = \alpha + \phi_1 x_{t-1} + \phi_2 x_{t-2} + ... + \phi_p x_{t-q} - \theta_1 \varepsilon_{t-1} - \theta_2 \varepsilon_{t-2} - ... - \theta_q \varepsilon_{t-q} + \varepsilon_t$$

Here, Φ and θ represent AR and MA coefficients. The α and ε_t captures the intercept and error at time t. The form gets very complicated as p and q increase; thus, lag operators are utilized for a concise representation of ARMA models. Let's assume that L represents the lag operator and, depending on the unit moved, we apply it by k times. These operators are also referred to as back shift operators.

$$Lx_t = x_{t-1}$$

$$L^2 x_t = x_{t-2}$$

$$...$$

$$L^p x_t = x_{t-p}$$

Using a lag operator, we can rewrite first-order auto-regressive models as follows:

$$AR(1): (1 - \phi L)x_t = \varepsilon_t$$

Similarly, the moving order first order equation can be written as follow:

$$MA(1): x_t = (1 - \theta L)\varepsilon_t$$

The preceding equations can be extended for higher-order AR and MA models:

$$AR(p): (1 - \phi_1 L - \phi_2 L^2 ... - \phi_p L^p)x_t = \varepsilon_t$$

The preceding two can be combined to form ARMA represented as follows:

$$MA(q): x_t = (1 - \theta_1 L - \theta_2 L^2 ... - \theta_q L^q)\varepsilon_t$$

The preceding representations are also used to study impulse-response functions. The impulse-response function captures the effect on response for x_t given a shock at time l where $l < t$. The impulse -esponse can also be considered as an effect on the dynamic system response given some external change.

Let's generate an ARMA(1,1) dataset by updating the script used previously with updated `ar` and `ma` components. We will also restrict the number of samples to 600 for simplicity:

```
# Number of samples
n = 600
# Generate AR(1) dataset
ar = np.r_[1, 0.6]
ma = np.r_[1, 0.3]
ar1ma1_data = smtsa.arma_generate_sample(ar=ar, ma=ma, nsample=n)
plotds(ar1ma1_data )
```

The signal and ACF and PACF plots are shown here:

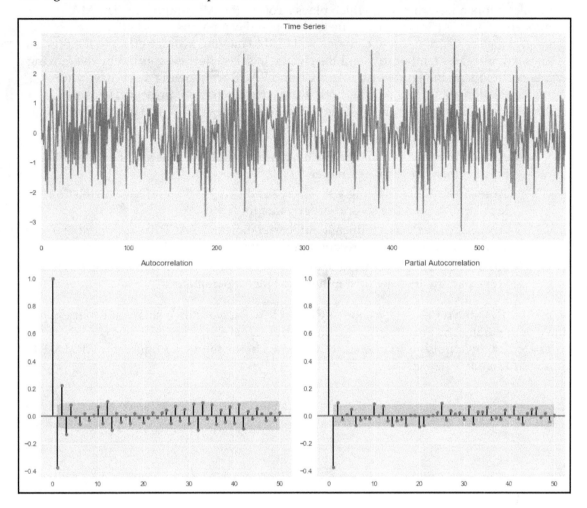

Figure 4.14: ARMA(1,1) signal with ACF and PACF

Sales processes in general follow an ARMA(1,1) model as sales in time *t* is a function of prior sales happening in time *t-1*, which plays a role in the AR component. The MA component of ARMA(1,1) is caused due to time-based campaigns launched by the company, such as distribution of coupons will lead to moving average effect to the process as sales will increases temporarily and the change in sales effect is captured by the moving average component. In Figure 4.14, both ACF and PACF are showing a sin curve with strong correlation at initial lags; thus, both p and q parameters are present. There are multiple scenarios to select p and q; some of the thumb rules that can be used to determine the order of ARMA components are as follows:

- Autocorrelation is exponentially decreasing and PACF has significant correlation at lag 1, then use the p parameter
- Autocorrelation is forming a sine-wave and PACF has significant correlation at lags 1 and 2, then use second-order value for p
- Autocorrelation has significant autocorrelation and PACF has exponential decay, then moving average is present and the q parameter needs to be set up
- Autocorrelation shows significant serial correlation and the PACF shows sine-wave pattern, then set up a moving average q parameter

In ARMA(1,1) time series data, as both ACF and PACF have shown sine-wave pattern, p and q both parameters are affecting the time series signal. The impact of lags can be computed using impulse-response curve, as shown in the following figure, for ARMA(1,1) time series signal:

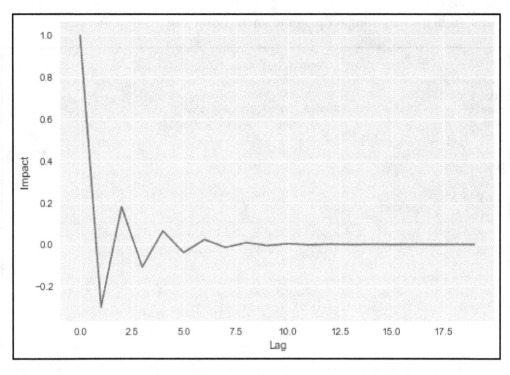

Figure 4.15: Impulse-response curve for ARMA(1,1) dataset

Figure 4.15 shows that, after five lags, there is very minimal effect on response. To evaluate the *AR* and *MA* values from the data, the `ARMA.fit` function from the `statsmodels.tsa.api` module is utilized, as shown in the following script:

```
# Build AR(1) model
ar1ma1 = smtsa.ARMA(ar1ma1_data.tolist(), order=(1, 1)).fit(
    maxlag=30, method='mle', trend='nc')
ar1ma1.summary()
```

The output from the preceding script is shown in Figure 4.16:

```
***
                        ARMA Model Results
==================================================================================
Dep. Variable:                      y    No. Observations:            600
Model:                     ARMA(1, 1)    Log Likelihood           -811.950
Method:                           mle    S.D. of innovations         0.936
Date:               Mon, 21 Aug 2017    AIC                       1629.900
Time:                        05:56:38    BIC                       1643.091
Sample:                             0    HQIC                      1635.035

===================================================================================
                 coef    std err          z      P>|z|      [0.025      0.975]
-----------------------------------------------------------------------------------
ar.L1.y       -0.5823      0.095     -6.100      0.000      -0.769      -0.395
ma.L1.y        0.2972      0.112      2.653      0.008       0.078       0.517
                                     Roots
===================================================================================
                  Real           Imaginary           Modulus          Frequency
-----------------------------------------------------------------------------------
AR.1           -1.7173            +0.0000j            1.7173             0.5000
MA.1           -3.3643            +0.0000j            3.3643             0.5000
-----------------------------------------------------------------------------------
***
```

Figure 4.16: Impulse-response curve for ARMA(1,1)

The model outcome shows an AR coefficient value of 0.58 and MA value of 0.29, which is close to values 0.6 and 0.3 used by the AR and MA component, respectively, to generate the time series signal. Also, alkaline information criteria (AIC) is another indicator used to evaluate model performance with an objective to minimize AIC. To set up a data-driven evaluation for p and q orders, AIC can be used as a criteria. An illustration of AIC minimization on ARMA(1,1) dataset is shown here:

```
# Optimize ARMA parameters
aicVal=[]
for ari in range(1, 3):
    for maj in range(1,3):
        arma_obj = smtsa.ARMA(ar1ma1_data.tolist(), order=(ari,
maj)).fit(maxlag=30, method='mle', trend='nc')
        aicVal.append([ari, maj, arma_obj.aic])
```

The output from the preceding script is summarized using p and q as model input and AIC as model output criteria:

S. No.	AR(p)	MA(q)	AIC
1	1	1	1629.8
2	1	2	1631.8
3	2	1	1631.8
4	2	2	1632.8

Table 4.17: AIC values for different p and q valuesof an ARMA model

Table 4.17 shows that ARMA(1,1) is the most optimal model with minimum AIC value; however, the difference between ARMA(1,1) and ARMA(2,2) is not high as ARMA(1,1) is a simpler model with less **degree of freedom (DOF)**; thus, ARMA(1,1) will be preferred over other complex models.

Let's illustrate an ARMA model using real-time series data. The dataset selected for illustration is IBM stock prices data from 1962 to 1965. The first step is to load the required modules and dataset into the Python environment:

```
# Load modules
from __future__ import print_function
import os
import pandas as pd
import numpy as np
from matplotlib import pyplot as plt

# Load Dataset
ibm_df = pd.read_csv('datasets/ibm-common-stock-closing-prices.csv')
ibm_df.head()

#Rename the second column
ibm_df.rename(columns={'IBM common stock closing prices': 'Close_Price'},
inplace=True)
ibm_df.head()
```

The preceding script utilizes pandas to load the dataset. The column names are renamed using the `rename` function supported by pandas DataFrame. The IBM stock closing price dataset looks as follows:

Figure 4.18: IBM stock dataset

The IBM plot shows a significant trend in the data over time. The next step in the process is to look at an autocorrelation plot for the dataset. The ACF and PACF plots can be obtained using the following script:

```
from statsmodels.graphics.tsaplots import import plot_acf, plot_pacf
ibm_df['residual']=ibm_df['Close_Price']-ibm_df['Close_Price'].mean()
ibm_df=ibm_df.dropna()
plot_acf(ibm_df.residual, lags=50)
plot_pacf(ibm_df.residual, lags=50)
```

The ACF and PACF plots produced by the preceding script are shown in Figure 4.16:

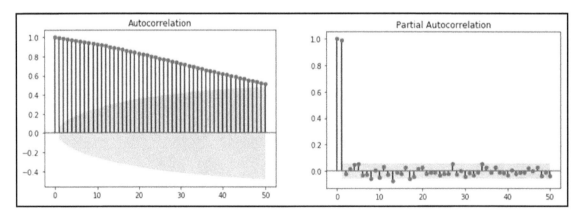

Figure 4.19: ACF and PACF for residual from the IBM stock dataset

The ACF is linearly decaying, showing a strong serial correlation; however, the partial autocorrelation shows only one step dependency. Also, as the chart shows a positive autocorrelation, correction should be done using the AR component with first-order correlation. The QQ-plot for the signal helps you evaluate the normality assumption. The QQ-plot for the IBM stock dataset is shown in Figure 4.20:

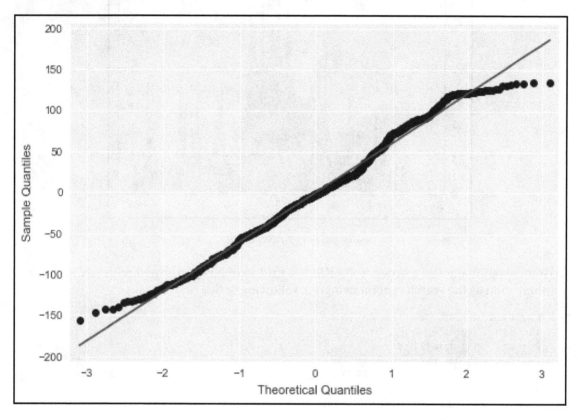

Figure 4.20: QQ plot for residual

The residual plot in Figure 4.21 shows that the dataset is close to bell curves with corner cases not following normal distribution, as shown in Figure 4.18, as a histogram plot:

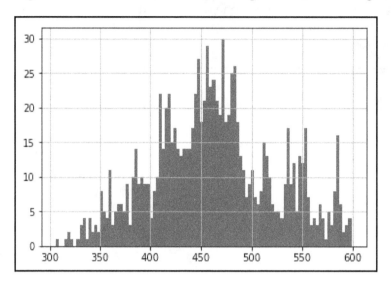

Figure 4.21: Histogram plot of IBM stock closing price

To get the optimal p and q orders for ARMA, a grid search is performed with AIC minimization as the search criteria using the following script:

```
# Optimize ARMA parameters
aicVal=[]
for ari in range(1, 3):
    for maj in range(0,3):
        arma_obj = smtsa.ARMA(ibm_df.Close_Price.tolist(), order=(ari,
maj)).fit(maxlag=30, method='mle', trend='nc')
        aicVal.append([ari, maj, arma_obj.aic])
```

The ARMA.fit function is used to fit the ARMA forecasting model with defined p and q models using maximum likelihood criteria. The outcome from the model is saved in an *aicVal* list, as shown in Figure 4.22:

S. No.	AR(p)	MA(q)	AIC
1	1	0	6702.7
2	1	1	6704.7
3	1	2	6706.6
4	2	0	6704.7
5	2	1	6705.7
6	2	2	6707.7

Figure 4.22: AIC score of different order of ARMA model

The AIC recommends the ARMA(1,1) model as the optimal model with minimum AIC value. The ARMA(1,1) model is refitted as the optimal model using the following script:

```
# Building optimized model using minimum AIC
arma_obj_fin = smtsa.ARMA(ibm_df.Close_Price.tolist(), order=(1,
1)).fit(maxlag=30, method='mle', trend='nc')
ibm_df['ARMA']=arma_obj_fin.predict()
```

The AIC score recommends a model of ARMA(1,0) with AIC score of 6702.7. The actual versus fitted value using ARMA(1,0) is as follows:

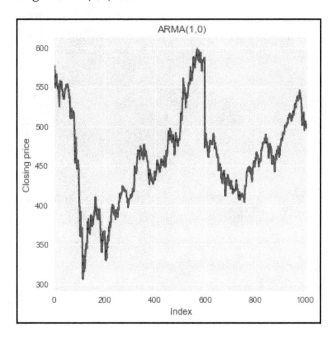

Figure 4.23 Actual versus forecasted value

The preceding plot shows a good fitting of the ARMA(1,0) model to predict stock closing prices:

```
# Plot the curves
f, axarr = plt.subplots(1, sharex=True)
f.set_size_inches(5.5, 5.5)
ibm_df['Close_Price'].iloc[1:].plot(color='b', linestyle = '-', ax=axarr)
ibm_df['ARMA'].iloc[1:].plot(color='r', linestyle = '--', ax=axarr)
axarr.set_title('ARMA(1,0)')
plt.xlabel('Index')
plt.ylabel('Closing price') Plot the curves
f, axarr = plt.subplots(1, sharex=True)
f.set_size_inches(5.5, 5.5)
ibm_df['Close_Price'].iloc[1:].plot(color='b', linestyle = '-', ax=axarr)
ibm_df['ARMA'].iloc[1:].plot(color='r', linestyle = '--', ax=axarr)
axarr.set_title('ARMA(1,1)')
plt.xlabel('Index')
plt.ylabel('Closing price')
```

One of the major limitations of these models are that they ignore the volatility factor making the signal non-stationary. The AR modeling is under consideration process is stationary, that is, error term is IID and follows normal distribution $\varepsilon t \sim N(0, \sigma^2_\varepsilon)$ and $|\Phi| < 1$. The $|\Phi| < 1$ condition makes the time series a finite time series as the effect of more recent observations in time series would be higher as compared to prior observations. The series that do not satisfy these assumptions fall into non-stationary series. An example of the non-stationary process is shown here:

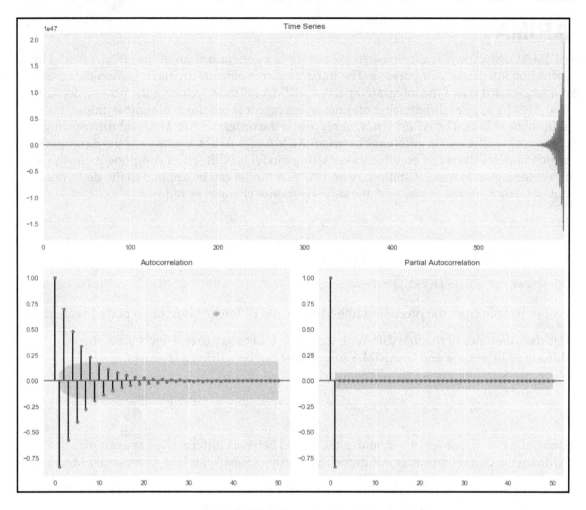

Figure 4.24: AR(1) non-stationary process

From the preceding plot, it can be seen that the variance of the process keeps increasing on the end of the dataset and there is a strong trend that is observed in ACF. The ARIMA discussed in the next section takes into account scenarios that are non-stationary for the forecasting.

ARIMA

ARIMA, also known as the Box-Jenkins model, is a generalization of the ARMA model by including integrated components. The integrated components are useful when data has non-stationarity, and the integrated part of ARIMA helps in reducing the non-stationarity. The ARIMA applies differencing on time series one or more times to remove non-stationarity effect. The ARIMA(p, d, q) represent the order for AR, MA, and differencing components. The major difference between ARMA and ARIMA models is the d component, which updates the series on which forecasting model is built. The d component aims to de-trend the signal to make it stationary and ARMA model can be applied to the de-trended dataset. For different values of d, the series response changes as follows:

For d=0: $x_t = x_t$

For d=1: $x_t = x_t - x_{t-1}$

For d=2: $x_t = (x_t - x_{t-1}) - (x_{t-1} - x_{t-2}) = x_t - 2x_{t-1} - x_{t-2}$

As can be seen from the preceding lines, the second difference is not two periods ago, rather it is the difference of the first different, that is, d=1. Let's say that $\hat{x}_t$ represents the differenced response and so ARIMA forecasting can be written as follows:

$$\hat{x}_t = \phi_1 \hat{x}_{t-1} + \phi_2 \hat{x}_{t-2} + \ldots + \phi_p \hat{x}_{t-q} + \theta_1 \varepsilon_{t-1} + \theta_2 \varepsilon_{t-2} + \theta_q \varepsilon_{t-q} + \varepsilon_t$$

Depending on the order of p, d, and q, the model behaves differently. For example, ARIMA(1,0, 0) is a first-order AR model. Similarly, ARIMA(0,0,1) is a first-order MA model.

Let's take an example of ARIMA(0,1,0) to illustrate the different components of ARIMA modeling. The ARIMA(0,1,0) represents a random walk model. The random walk model depends only on the last time instance and can be represented as follows:

$$x_t = x_{t-1} + \varepsilon_t$$

The preceding random walk equation can be also represented in lag operators as follows:

$$(1-L)x_t = \varepsilon_t$$

Here, $\varepsilon t \sim N(0,\sigma^2)$ is the error component and follows normal distribution. Adding a constant to the preceding random walk model will cause a drift in the model, which is also stochastic in nature as in the following equation:

$$(1-L)x_t = \alpha + \varepsilon_t$$

Here, α is the drift operator that will give a drifting effect to the time series signals. Let's illustrate ARIMA modeling using Dow Jones Indices time series dataset (DJIA) from 2016. The DJIA dataset with its basic ACF and PACF plots are shown in Figure 4.25:

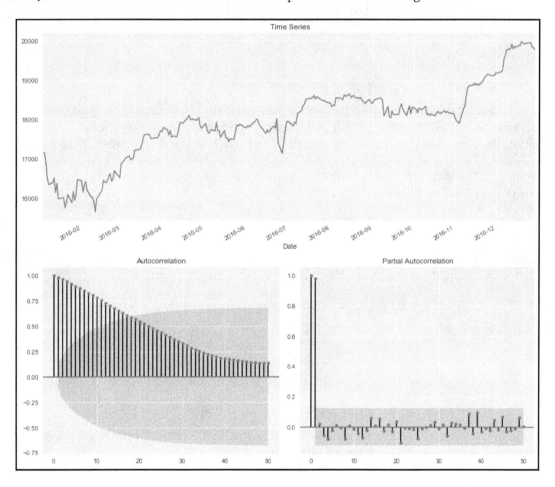

Figure 4.25: DJIA dataset with its ACF and PACF

The dataset clearly shows a non-stationary signal with an increasing trend. The ACF is also showing an exponential decay while PACF has a strong correlation in lag 2. The non-stationarity can also be checked by evaluating mean and variance in different time periods. The difference in mean and variance validates the hypothesis of non-stationarity. For example, we split the DJIA dataset into two semesters from Jan-June 2016 and July-December 2016 and evaluate the mean and variance as follows:

```
mean1, mean2 =djia_df.iloc[:125].Close.mean(),
djia_df.iloc[125:].Close.mean()
var1, var2 = djia_df.iloc[:125].Close.var(), djia_df.iloc[125:].Close.var()
print('mean1=%f, mean2=%f' % (mean1, mean2))
print('variance1=%f, variance2=%f' % (var1, var2))

# Output
mean1=17226.579164, mean2=18616.603593
variance1=487045.734003, variance2=325183.639530
```

The evaluated mean and variance for both semesters show significant difference in mean and variance values, thus suggesting that the data is non-stationary. Another way to evaluate the non-stationarity is using statistical tests such as the Augmented Dickey-Fuller (ADF) test. The ADF is a unit root test that evaluates the strength of trend in a time series component. The ADF uses higher-order AR models with an objective to optimize information criterion. Suppose for an AR(3) model:

$$x_t - \phi_1 x_{t-1} - \phi_2 x_{t-2} - \phi_3 x_{t-3} = \varepsilon_t$$

The preceding equation can be rewritten using a difference lag term:

$$x_t = (\phi_1 + \phi_2 + \phi_3)x_{t-1} - (\phi_2 + \phi_3)(x_{t-1} - x_{t-2}) - \phi_3(x_{t-2} - x_{t-3}) + \varepsilon_t$$

Let's reduce the preceding equation as follows:

$$x_t = \rho_1 x_{t-1} - \rho_2(x_{t-1} - x_{t-2}) - \rho_3(x_{t-2} - x_{t-3}) + \varepsilon_t$$

The preceding equation can be rewritten using a lag operator:

$$(1 - \phi_1 L - \phi_2 L^2 - \phi_3 L^3)x_t = \varepsilon_t$$

The ADF solves the preceding equation assuming that L=1 is the solution of the preceding polynomial, which can be represented as follows:

$$1 - \phi_1 L - \phi_2 L^2 - \phi_3 L^3 = 0$$

Putting L=1, the preceding equation can be reduced to the following:

$$\phi_1 + \phi_2 + \phi_3 = 1$$

For unit root, the lag difference can be rewritten as follows:

$$\Delta x_t = (\rho_1 - 1)x_{t-1} + \sum_{i=2} \rho_i (\Delta x_{t-i+1}) + \varepsilon_t$$

The preceding equation is used to determine the AR lag component using Schwartz Bayesian information criterion or minimizing the **Akaike information criterion (AIC)**. In the presence of strong auto correlation, the original series needs differencing. The NULL hypothesis of the ADF proposes that $H_0 : \varrho = 0$ against the alternative hypothesis $H_0 : \varrho < 0$. In other words, the null hypothesis is the presence of the unit root or non-stationarity whereas the alternate hypothesis suggests stationarity of the data.

Let's run the ADF test for the DJIA dataset:

```
# ADF Test
from statsmodels.tsa.stattools import adfuller
adf_result= adfuller(djia_df.Close.tolist())
print('ADF Statistic: %f' % adf_result[0])
print('p-value: %f' % adf_result[1])
```

The output for ADF is shown here:

```
In [441]: from statsmodels.tsa.stattools import adfuller
     ...: adf_result= adfuller(djia_df.Close.tolist())
     ...: print('ADF Statistic: %f' % adf_result[0])
     ...: print('p-value: %f' % adf_result[1])
ADF Statistic: -0.462320
p-value: 0.899162

In [442]:
```

Figure 4.26: Output from the ADF test

Ideally, a more negative value of ADF statistics will represent a stationary signal. For the given dataset, as *p-value* is quite high, we cannot reject the NULL hypothesis making it a non-stationary signal. Most of the packages ensure stationarity is satisfied before executing the models. The *qqplot* to visualize normality for the DJIA dataset is shown in Figure 4.27:

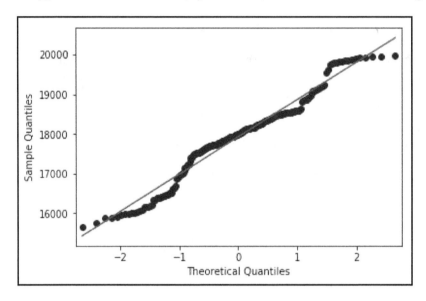

Figure 4.27: *qqplot* for the DJIA dataset

Figure 4.27 shows significant non-normality within the DJIA dataset. Most of the core packages written in Python check for the stationarity within a dataset. In case stationarity is not satisfied, an error is raised. Let's try to force fit an ARMA(1,1) model on the current dataset:

```
# Force fit ARMA(1,1) model on non-stationary signal
arma_obj = smtsa.ARMA(djia_df['Close'].tolist(), order=(1,
1)).fit(maxlag=30, method='mle', trend='nc')
```

The preceding force fitting will raise the following error:

```
File "<ipython-input-442-50da6eb526c0>", line 1, in <module>
    arma_obj = smtsa.ARMA(djia_df['Close'].tolist(), order=(1, 1)).fit(maxlag=30,
method='mle', trend='nc')

File "C:\ProgramData\Anaconda3\lib\site-packages\statsmodels\tsa\arima_model.py", line
956, in fit
    start_ar_lags)

File "C:\ProgramData\Anaconda3\lib\site-packages\statsmodels\tsa\arima_model.py", line
574, in _fit_start_params
    start_params = self._fit_start_params_hr(order, start_ar_lags)

File "C:\ProgramData\Anaconda3\lib\site-packages\statsmodels\tsa\arima_model.py", line
557, in _fit_start_params_hr
    raise ValueError("The computed initial AR coefficients are not "

ValueError: The computed initial AR coefficients are not stationary
You should induce stationarity, choose a different model order, or you can
pass your own start_params.
```

Figure 4.28: Output when the ARMA model is fitted on a non-stationary signal

Differencing will help make the signal stationary:

```
#Let us plot the original time series and first-differences
first_order_diff = djia_df['Close'].diff(1)
fig, ax = plt.subplots(2, sharex=True)
fig.set_size_inches(5.5, 5.5)
djia_df['Close'].plot(ax=ax[0], color='b')
ax[0].set_title('Close values of DJIA during Jan 2016-Dec 2016')
first_order_diff.plot(ax=ax[1], color='r')
ax[1].set_title('First-order differences of DJIA during Jan 2016-Dec 2016')
```

The output from the preceding script is shown in Figure 4.29:

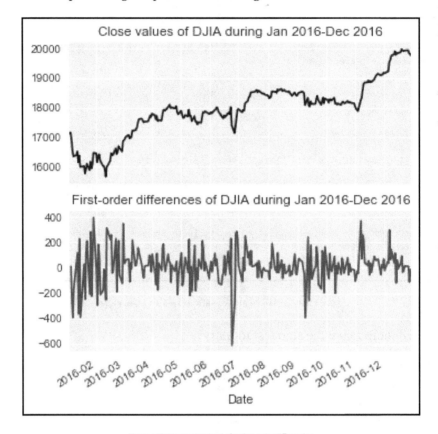

Figure 4.29: Time series signal after first-order differencing

The ACF and PACF chart for an integrated signal with *d*=1 is shown in Figure 4.30:

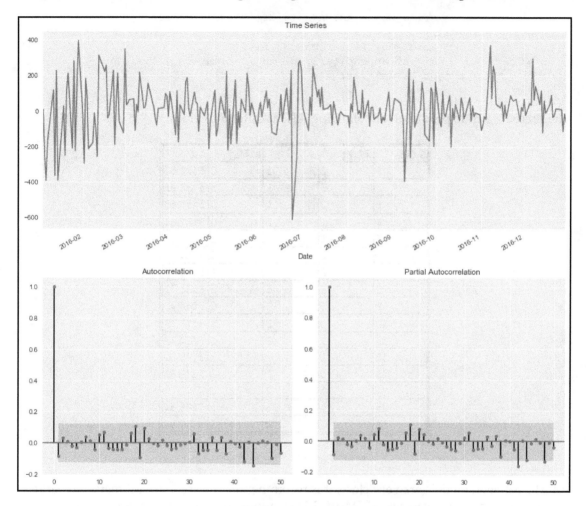

Figure 4.30: First-order difference ACF and PACF plots for the DJIA dataset

The ADF statistics on residual have a value of -17.13 with *p-val* close to zero, thus stating that the model is stationary; however, both ACF and PACF show not much of trend of the moving average component showing a random walk behavior. Also, another way to run is to optimize using AIC as a criteria:

```
# Optimize ARMA parameters
aicVal=[]
for d in range(1,3):
    for ari in range(0, 3):
```

```
for maj in range(0,3):
    try:
        arima_obj = ARIMA(djia_df['Close'].tolist(),
order=(ari,d,maj))
        arima_obj_fit=arima_obj.fit()
        aicVal.append([ari, d, maj, arima_obj_fit.aic])
    except ValueError:
        pass
```

The preceding code produces the following AIC at different values:

S. No.	AR(P)	d	MA(q)	AIC
1	0	1	0	3182.64
2	0	1	1	3182.93
3	0	1	2	3184.66
4	1	1	0	3182.84
5	1	1	1	3184.74
6	2	1	0	3184.70
7	2	1	1	3186.77
8	2	1	2	3188.71
9	0	2	0	3364.45
10	0	2	1	3177.77
11	1	2	0	3274.73
12	1	2	1	3178.00
13	1	2	2	3179.89
14	2	2	0	3242.25
15	2	2	1	3179.84
16	2	2	2	3180.94

Figure 4.31: AIC values for different ARIMA model

As AIC between models are very close, it is recommended to use a subject matter expert to pick the right model. Let's pick ARIMA(0,2,1) for model fitting and evaluation. The ARIMA(0,2,1) applies second-order differencing and first-order moving average component to determine the relationship between observations. The model parameter is set up as shown in the following script:

```
# Evaluating fit using optimal parameter
arima_obj = ARIMA(djia_df['Close'].tolist(), order=(0,2,1))
arima_obj_fit = arima_obj.fit(disp=0)
arima_obj_fit.summary()

# Evaluate prediction
pred=np.append([0,0],arima_obj_fit.fittedvalues.tolist())
```

```
djia_df['ARIMA']=pred
diffval=np.append([0,0], arima_obj_fit.resid+arima_obj_fit.fittedvalues)
djia_df['diffval']=diffval
```

The comparison with the actual and forecasted values is obtained and visualized using the following script:

```
# Plot the curves
f, axarr = plt.subplots(1, sharex=True)
f.set_size_inches(5.5, 5.5)
djia_df['diffval'].iloc[2:].plot(color='b', linestyle = '-', ax=axarr)
djia_df['ARIMA'].iloc[2:].plot(color='r', linestyle = '--', ax=axarr)
axarr.set_title('ARIMA(0,2,1)')
plt.xlabel('Index')
plt.ylabel('Closing price')
```

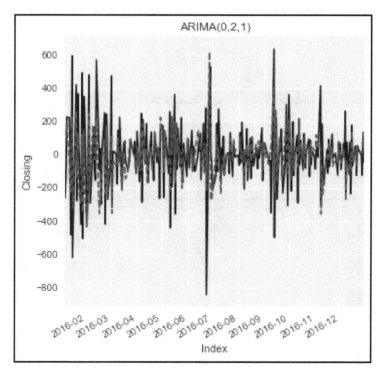

Figure 4.32: Forecasted output from the ARIMA (0,2,1) model

>

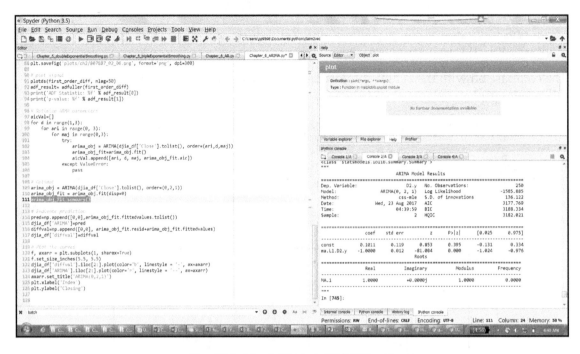

Figure 4.33: Coefficients for MA using the ARIMA(0,2,1) model

Another critical check of whether the model is optimally built is white noise, such as the behavior of an error can be observed using *qqplot*, as shown in Figure 4.34:

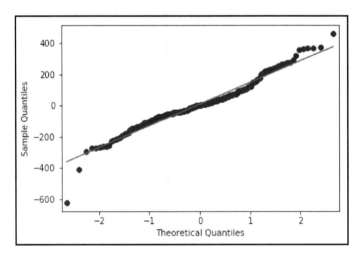

Figure 4.34: QQ normality plot for residual from the ARIMA(0,2,1) model

The QQ-normality plot with the ARIMA(0,2,1) model shows significant normality fit, as shown in Figure 4.27. The error normality can also be evaluated using the Shapiro-wilk test.

The further extension of the ARIMA model includes the seasonality component for AR, I, and MA represented in capitals. The seasonal ARIMA is represented as ARIMA(p,d,q) (P, D, Q)$_m$, where P, D, and Q represent the seasonal part of auto-regressive, integrated, and moving average, respectively. The m in the seasonal ARIMA model represents the number of periods per season. In scenarios when seasonality is present, the extra steps of seasonal difference and seasonal adjustment may be required to ensure that the signal is stationary. For example, if you look at DJIA difference ACF and PACF plots shown in Figure 4.30, the autocorrelation becomes slightly significant at 42 index, which means that there may be seasonality present. The seasonality present on first difference and can be seen using the following lines of code:

```
# Seasonality (based on first difference ACF shows significance at 42 lag)
x=djia_df['Close']-djia_df['Close'].shift(42)
x.plot()
```

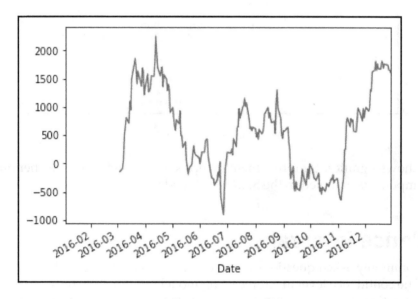

Figure 4.35: Seasonality in the DJIA dataset

The preceding seasonality can be corrected using seasonal ARIMA supported in the *statmodels.SARIMAX* model. The script to set up a seasonal ARIMA model for the DJIA dataset is as follows:

```
# Seasonality (based on first difference ACF shows significance at 42 lag)
x=djia_df['Close']-djia_df['Close'].shift(42)
```

```
mod = sm.tsa.statespace.SARIMAX(djia_df['Close'], trend='n', order=(0,2,1),
seasonal_order=(1,1,1,42))
sarimax= mod.fit()
sarimax.summary()
```

The following is the outcome of the preceding script:

```
Out[55]:
<class 'statsmodels.iolib.summary.Summary'>
"""
                         Statespace Model Results
==============================================================================
Dep. Variable:                      Close   No. Observations:             252
Model:           SARIMAX(0, 2, 1)x(1, 1, 1, 42)   Log Likelihood     -1366.570
Date:                     Sun, 27 Aug 2017   AIC                      2741.141
Time:                             17:32:02   BIC                      2755.258
Sample:                           01-04-2016   HQIC                   2746.821
                               - 12-30-2016
Covariance Type:                      opg
==============================================================================
                 coef    std err          z      P>|z|      [0.025      0.975]
------------------------------------------------------------------------------
ma.L1         -1.0000     16.149     -0.062      0.951     -32.651      30.651
ar.S.L42      -0.2542      0.093     -2.727      0.006      -0.437      -0.072
ma.S.L42      -0.4671      0.130     -3.597      0.000      -0.722      -0.213
sigma2       2.355e+04    3.8e+05      0.062      0.951    -7.21e+05    7.68e+05
==============================================================================
Ljung-Box (Q):                       31.80   Jarque-Bera (JB):           21.28
Prob(Q):                              0.82   Prob(JB):                    0.00
Heteroskedasticity (H):               0.41   Skew:                       -0.24
Prob(H) (two-sided):                  0.00   Kurtosis:                    4.49
==============================================================================
```

Figure 4.36: Output from the SARIMAX model

The model shows significant improvement in terms of AIC and can be further optimized for different components involved in the SARIMAX model.

Confidence interval

One of the commonly asked questions in forecasting is, What is the confidence interval for estimates? The confidence level in a forecasting model is defined by the *alpha* parameter in the forecast function. The *alpha* value 0.05 represents an estimate with 95% confidence, which can be interpreted as the estimates returned by the model have 5% probability of not falling in the defined distribution range. The confidence interval is evaluated as follows:

$$\hat{x}_t \pm Z_\alpha \frac{\sigma}{\sqrt{N}}$$

Here, Z_a is the critical value defined based on alpha. For the *alpha* value 0.05, the critical value is 1.96. The confidence interval with an alpha value 0.05 for the DJIA dataset modeled using the ARIMA(0,2,1) model can be obtained using the *forecast* function from the `arima_obj_fit` object:

```
# Forecasting and CI
f, err, ci=arima_obj_fit.forecast(40)
plt.plot(f)
plt.plot(ci)
plt.xlabel('Forecasting Index')
plt.ylabel('Forecasted value')
```

The forecasting estimates and confidence interval obtained using the preceding script is shown in Figure 4.37:

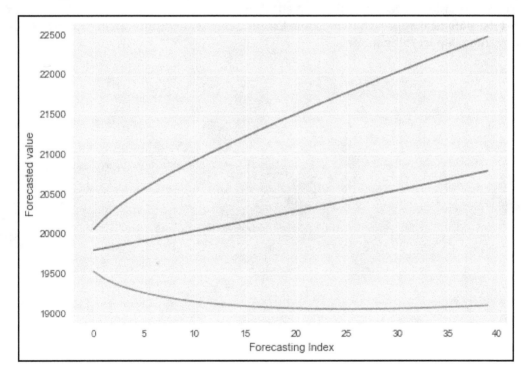

Figure 4.37: Forecast and confidence interval with 95% confidence

Summary

In this chapter, we covered auto-regressive models such as a MA model to capture serial correlation using error relationship. On similar lines, AR models were covered, which set up the forecasting using the lags as dependent observations. The AR models are good to capture trend information. The ARMA-based approach was also illustrated, which integrates AR and MA models to capture any time-based trends and catastrophic events leading to a lot of error that will take time to correct such as an economy meltdown. All these models assume stationarity; in scenarios where stationarity is not present, a differencing-based model such as ARIMA is proposed, which performs differencing in time series datasets to remove any trend-related components. The forecasting approaches were illustrated with examples using Python's *tsa* module.

The current chapter focuses on using statistical methods for forecasting. The next chapter extend the statistical approach to machine learning methods for forecasting specifically looking into deep learning models.

5

Deep Learning for Time Series Forecasting

So far in this book, we have described traditional statistical methods for time series analysis. In the preceding chapters, we has discussed several methods to forecast the series at a future point in time from observations taken in the past. One such method to make predictions is the **auto-regressive (AR)** model, which expresses the series at time t as a linear regression of previous p observations:

$$x_t = w_0 + \sum_{i=1}^{p} w_i x_{t-i} + \epsilon_t$$

Here, ϵ_t is the residual error term from the AR model.

The idea underlying the linear model can be generalized that the objective of time series forecasting is to develop a function f that predicts x_t in terms of the observations at previous p points of time:

$$x_t = f(x_{t-1}, x_{t-2}, \dots , x_{t-p})$$

In this chapter, we will explore three methods based on neural networks to develop the function f. Each method includes defining a neural network architecture (in terms of the number of hidden layers, number of neurons in every hidden layer, and so on) and then training the network using the backpropagation algorithm or its variant that is appropriate for the network architecture being used.

The last few years have witnessed a resurgence of interest in neural networks. This has been possible due to the availability of abundant training data from digital media and cheaper access to GPU-driven parallel computing. These factors have enabled training neural networks having hundreds of thousands, and in some cases, millions of parameters. Neural networks of different architectures have been successfully applied to solve problems in computer vision, speech recognition, and natural language translation. Research and practice of designing and training neural networks in these areas is popularly known as **deep learning** that, as a name, is indicative of the many hidden layers used in these models.

Deep learning has given interesting neural architectures that are designed to address the special structural characteristics of data on image and language. For example, **convolutional neural networks** (**CNNs**) are designed to take advantage of two- or three-dimensional structures of images, while most language models use **recurrent neural networks** (**RNNs**) that support sequence and memory inherently found in spoken and written languages. These new developments have also been applied to areas where traditionally statistical machine learning has dominated. One such area is time series forecasting.

In this chapter, we will explore three different types of neural networks for time series forecasting. We start with **multi-layer perceptrons** (**MLP**). This will be followed by recurrent neural networks that suits the sequential arrangement of data points. Finally, we will cover convolutional neural networks, which are mostly used on images, but we will discuss how special forms of CNNs can be used for time series forecasting. These topics will be covered through an explanation of the network architecture and how it can be applied to time series forecasting. Code demonstrations show you how to use cutting-edge deep learning libraries to develop models of time series forecasting. Examples in this chapter are implemented using the Keras API for deep learning. Keras is a high-level API that allows defining different neural network architectures and training them using various of gradient-based optimizers. In the backend, Keras uses a low-level computational framework that is implemented in C, C++, and FORTRAN. Several such low-level frameworks are available open source. Keras supports the following three: TensorFlow, which was developed by Google and is the default backend of Keras, CNTK, an open-source framework from Microsoft, and Theano, which was originally developed at University of Montreal, Canada. Examples in this book use TensorFlow as the backend. Hence, to run the examples, you would need both Keras and TensorFlow installed.

So, without further ado, let's dive into the fascinating topic of time series forecasting by deep learning.

Multi-layer perceptrons

Multi-layer perceptrons (MLP) are the most basic forms of neural networks. An MLP consists of three components: an input layer, a bunch of hidden layers, and an output layer. An input layer represents a vector of regressors or input features, for example, observations from preceding p points in time $[x_{t-1}, x_{t-2}, \dots, x_{t-p}]$. The input features are fed to a hidden layer that has n neurons, each of which applies a linear transformation and a nonlinear activation to the input features. The output of a neuron is $g_i = h(\mathbf{w}_i\mathbf{x} + b_i)$, where $\mathbf{w}_i$ and b_i are the weights and bias of the linear transformation and h is a nonlinear activation function. The nonlinear activation function enables the neural network to model complex non-linearities of the underlying relations between the regressors and the target variable. Popularly, h is the sigmoid function, $\frac{1}{1-e^{-z}}$, that squashes any real number to the interval **[0,1]**. Due to this property, the sigmoid function is used to generate binary class probabilities and hence is commonly used in classification models. Another choice of a nonlinear activation function is the `tanh` function, $\frac{1-e^{-z}}{1+e^{-z}}$, which binds any real number to the interval **[-1,1]**. In some cases, h is an identity or a linear function.

In case of a single hidden layer neural network (as shown on the left in the following figure), the output from each neuron is passed to the output layer, which applies a linear transformation and an activation function to generate the prediction of the target variable, which in case of time series forecasting, is the predicted value of the series at the tth point in time. In an MLP, multiple hidden layers are stacked against each other. The neurons' output from one hidden layer is fed as input to the next hidden layer. The neurons in this hidden layer transform the input and pass to the next hidden layer. Finally, the last hidden layer feeds into the output layer:

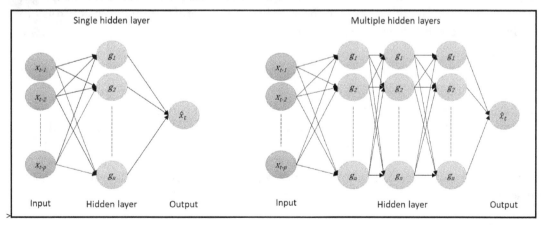

Figure 5.1: Multi-layer perceptron

The hidden layers of MLP are also called dense, or sometimes fully-connected, layers. The name dense bears significance to the fact that all neurons of a dense layer are connected to all neurons in the preceding and the following layer. If the preceding layer is the input layer, then all input features feed into each of the neurons of the hidden layer. Due to the many-to-many connections between the input layer and the first dense layer and between the dense layers themselves, an MLP has an enormous number of trainable weights. For example, if the number of input features is p and there are three dense layers having number of neurons n_1, n_2, and n_3 respectively, then the number of trainable weights is $p \times n_1 + n_1 \times n_2 + n_2 \times n_3 + n_3$. The last element in this calculation is the number of weights connecting the third hidden layer and the output layer. Deep MLPs have several dense layers and hundreds, even thousands, of neurons in each layer. Hence, the number of trainable weights in deep MLPs is very large.

Training MLPs

The weights **w** of a neural network are found by running a gradient based optimization algorithm such as stochastic gradient descent that iteratively minimizes the loss or error (**L**) incurred by the network in making predictions over the training data. **Mean-squared error (MSE)** and **mean absolute error (MAE)** (sometimes mean absolute percentage error) are frequently used for regression tasks while binary and categorical log loss are common loss functions for classification problems. For time series forecasting, MSE and MAE would be apt to train the neural models.

Gradient descent algorithms work by moving the weights, in iterations i, along their gradient path. The gradient is the partial derivative of the loss function L with respect to the weight. The simplest update rule to change a weight w requires the values of the weights, partial derivative of L with respect to the weights, and a learning rate α that controls how fast the point descends along the gradient:

$$w_{i+1} = w_i - \alpha \left(\frac{\partial L}{\partial w} \right)_{w=w_i}$$

This basic update rule has several variants that impacts the convergence of the algorithm. However, a crucial input to all gradient-based algorithms is the partial derivative that must be calculated for all weights of the network. In deep neural networks, some of which have millions of weights, the derivative calculations can be a behemoth computational task. This is exactly where the famous backpropagation algorithm comes in to solve this problem efficiently.

To understand backpropagation, one should first know computational graphs and how they are used to make computations in a neural network.

Let's consider a simple single hidden layer neural network having two hidden units each with a sigmoid activation. The output unit is a linear transformation of its inputs. The network is fed with two input variables, $[x_1, x_2]$. The weights are shown along the edges of the network:

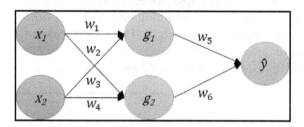

The network performs a series of additions, multiplications, and a couple of sigmoid functions to transform the input into a prediction $\hat{y}$. The transformation of input into a prediction is referred to as *forward pass* over the neural network. The following figure shows you how a forward pass is achieved by a computational graph for an input pair *[-1,2]*. Each computation results in an intermediate output p_i. Intermediate results p_7 and p_8 are the output of the hidden neurons g_1 and g_2. During training, the loss L is also computed in the forward pass.

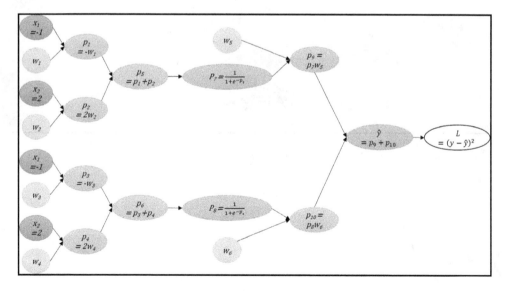

Figure 5.3: Computational graph of a single layer perceptron having two hidden neurons

At this point, the backpropagation algorithm is applied to compute the partial derivations between two nodes connected by an edge. The backward traversal in the graph to compute the partial derivative is also known as *backward pass*. The partial differentiation operator is applied at every node and the partial derivatives are assigned to the respective edges connecting the downstream node along the computational graph. Following the Chain Rule, the partial derivative $\frac{\partial L}{\partial w}$ is computed by multiplying the partial derivatives on all edges connecting the weight node and loss node. If multiple paths exist between a weight node and loss node, the partial derivatives along each path are added to get the total partial derivative of the loss with respect to the weight. This graph-based technique of implementing the *forward* and *backward* passes is the underlying computational trick used in powerful deep learning libraries. The *backward pass* is illustrated in the following figure:

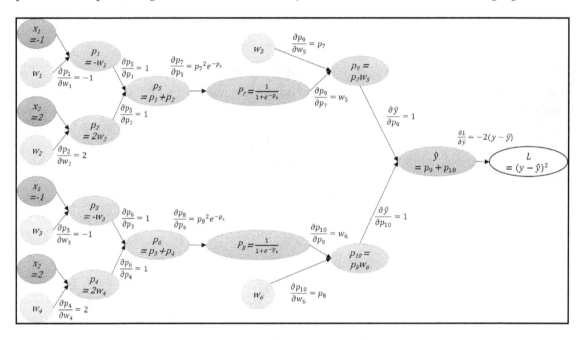

Figure 5.4: Calculation of partial derivatives in a computational graph

The partial derivatives of the loss with respect to the weights are obtained by applying the Chain Rule:

$$\frac{\partial L}{\partial w_5} = \frac{\partial L}{\partial \hat{y}} \frac{\partial \hat{y}}{\partial p_9} \frac{\partial p_9}{\partial w_5} = -2(y - \hat{y}) \times 1 \times p_7$$

$$\frac{\partial L}{\partial w_6} = \frac{\partial L}{\partial \hat{y}} \frac{\partial \hat{y}}{\partial p_{10}} \frac{\partial p_{10}}{\partial w_6} = -2(y - \hat{y}) \times 1 \times p_8$$

$$\frac{\partial L}{\partial w_1} = \frac{\partial L}{\partial \hat{y}} \frac{\partial \hat{y}}{\partial p_9} \frac{\partial p_9}{\partial p_7} \frac{\partial p_7}{\partial p_5} \frac{\partial p_5}{\partial p_1} \frac{\partial p_1}{\partial w_1} = -2(y - \hat{y}) \times 1 \times w_5 \times p_7^2 e^{-p_5} \times 1 \times -1$$

$$\frac{\partial L}{\partial w_2} = \frac{\partial L}{\partial \hat{y}} \frac{\partial \hat{y}}{\partial p_9} \frac{\partial p_9}{\partial p_7} \frac{\partial p_7}{\partial p_5} \frac{\partial p_5}{\partial p_2} \frac{\partial p_2}{\partial w_2} = -2(y - \hat{y}) \times 1 \times w_5 \times p_7^2 e^{-p_5} \times 1 \times 2$$

$$\frac{\partial L}{\partial w_3} = \frac{\partial L}{\partial \hat{y}} \frac{\partial \hat{y}}{\partial p_{10}} \frac{\partial p_{10}}{\partial p_8} \frac{\partial p_8}{\partial p_6} \frac{\partial p_6}{\partial p_3} \frac{\partial p_3}{\partial w_3} = -2(y - \hat{y}) \times 1 \times w_6 \times p_8^2 e^{-p_6} \times 1 \times -1$$

$$\frac{\partial L}{\partial w_4} = \frac{\partial L}{\partial \hat{y}} \frac{\partial \hat{y}}{\partial p_{10}} \frac{\partial p_{10}}{\partial p_8} \frac{\partial p_8}{\partial p_6} \frac{\partial p_6}{\partial p_4} \frac{\partial p_4}{\partial w_4} = -2(y - \hat{y}) \times 1 \times w_6 \times p_8^2 e^{-p_6} \times 1 \times 2$$

During training, the weights are initialized with random numbers commonly sampled from a uniform distribution with upper and lower limits of [*-1,1*] or a normal distribution having mean at zero and unit variance. This random initialization schemes have a few variants that enhance the convergence of the optimization. In this case, let's assume that the weights are initialized from a uniform random distribution and therefore, w_1 = -0.33, w_2 = 0.57, w_3 = 0.02, w_4 = -0.01, w_5=0.07, and w_6 = 0.82. With these values, let's walk through the *forward* and *backward* passes over the computational graph. We update the previous figure with the values computed during *forward pass* in blue and the gradients calculated during *backward pass* in red. For this example, we set the actual value of the target variable as y = 1:

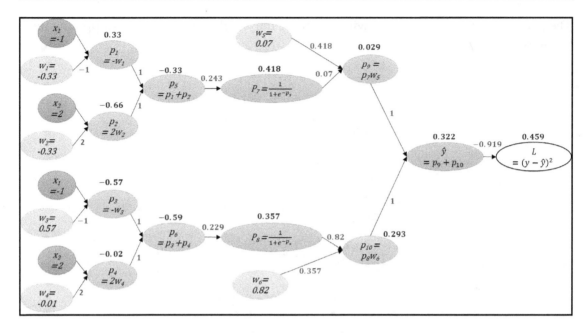

Figure 5.5: Forward (in blue) and backward (in red) passes over a computational graph

Once the gradients along the edges are computed, the partial derivatives with respect to the weights are just an application of the Chain Rule, which we discussed previously. The final values of the partial derivatives are as follows:

$$\frac{\partial L}{\partial w_5} = -0.919 \times 1 \times 0.418 = -0.384$$

$$\frac{\partial L}{\partial w_6} = -0.919 \times 1 \times 0.357 = -0.328$$

$$\frac{\partial L}{\partial w_1} = -0.919 \times 1 \times 0.07 \times 0.243 \times 1 \times -1 = 0.016$$

$$\frac{\partial L}{\partial w_2} = -0.919 \times 1 \times 0.07 \times 0.243 \times 1 \times 2 = -0.032$$

$$\frac{\partial L}{\partial w_3} = -0.919 \times 1 \times 0.82 \times 0.229 \times 1 \times -1 = 0.173$$

$$\frac{\partial L}{\partial w_4} = -0.919 \times 1 \times 0.82 \times 0.229 \times 1 \times 2 = -0.346$$

The next step is to update the weights using the gradient descent algorithm. Hence, with a learning rate of α = *0.01*, the new value of w_5 = 0.07 - 0.01 x -0.384 = 0.0738. The rest of the weights can also be updated using a similar update rule.

The process of iterative weight updates is repeated multiple times. The number of times the weight updates are done is known as number of epochs or passes over the training data. Usually, a tolerance criteria on the change of the loss function compared to the previous epoch controls the number of epochs.

The backpropagation algorithm along with a gradient-based optimizer is used to determine the weights of a neural network. Thankfully, there exist powerful deep learning libraries, such as Tensorflow, Theano and CNTK which implement computational graphs to train neural networks of any architecture and complexity. These libraries come with built-in support to run the computations as mathematical operations on multidimensional arrays and can also leverage GPUs to make faster calculations.

MLPs for time series forecasting

In this section, we will use MLPs to develop time series forecasting models. The dataset used for these examples is on air pollution measured by concentration of **particulate matter (PM)** of diameter less than or equal to 2.5 micrometers. There are other variables such as air pressure, air temperature, dew point, and so on. A couple of time series models have been developed-one on air pressure and the other on pm 2.5. The dataset has been downloaded from the UCI Machine Learning Repository. The link to the problem's description and datasets is `https://archive.ics.uci.edu/ml/datasets/Beijing+PM2.5+Data`.

The code for the time series model of air pressure is in the Jupyter notebook `code/Chapter_5_Air Pressure_Time_Series_Forecasting_by_MLP.ipynb` while the one on `pm2.5` is in `code/ Chapter_5_Air Pressure_Time_Series_Forecasting_by_MLP.ipynb`. The code folder is in the book's GitHub repository. Let's now describe how the time series model on air pressure has been developed.

We start by importing the packages required to run the code:

```
from __future__ import print_function
import os
import sys
import pandas as pd
import numpy as np
%matplotlib inline
```

```
from matplotlib import pyplot as plt
import seaborn as sns
import datetime
```

After setting the current working directory, the dataset is read from the `.csv` file to
`pandas.DataFrame`:

```
#set current working directory
os.chdir('D:/Practical Time Series')
#Read the dataset into a pandas.DataFrame
df = pd.read_csv('datasets/PRSA_data_2010.1.1-2014.12.31.csv')
```

To make sure that the rows are in the right order of date and time of observations, a new
column `datetime` is created from the date- and time-related columns. The new column
consists of Python's `datetime.datetime` objects. The `DataFrame` is sorted in ascending
order over this column:

```
df['datetime'] = df[['year', 'month', 'day', 'hour']]
                    .apply(lambda row: datetime.datetime(year=row['year'],
month=row['month'], day=row['day'],
hour=row['hour']), axis=1)
df.sort_values('datetime', ascending=True, inplace=True)
```

The `PRES` column has the data on air pressure. Let's draw a box plot to visualize the central
tendency and dispersion of `PRES`:

```
plt.figure(figsize=(5.5, 5.5))
g = sns.boxplot(df['PRES'])
g.set_title('Box plot of Air Pressure')
```

Though the box plot does not show the actual time series, it is useful in quickly identifying
the presence of outliers. By convention, observations falling outside 25th quartile – 1.5 times
of the inter-quartile range and 75th quartile + 1.5 times of the inter-quartile range are
designated as outliers. The inter-quartile range is 75th quartile – 25th quartile. Presence of
outliers can be used to decide the loss function for training the neural network as we will
see in the examples of this chapter.

The box-plot of air pressure shows that there are no observations that can be designated as outliers.

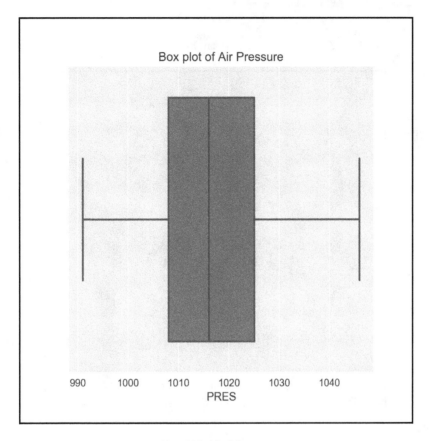

Figure 5.6: Boxplot of air pressure

Next, we plot the time series of air pressure for the entire duration of the observations:

```
g = sns.tsplot(df['PRES'])
g.set_title('Time series of Air Pressure')
g.set_xlabel('Index')
g.set_ylabel('Air Pressure readings in hPa')
```

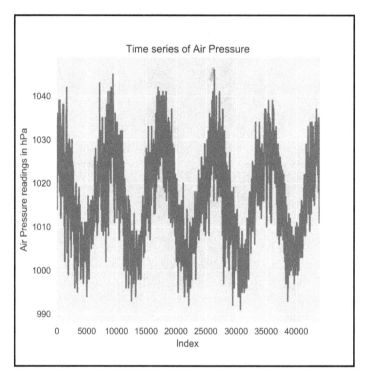

Figure 5.7: Time series of air pressure

Gradient descent algorithms perform better (for example, converge faster) if the variables are within range [-1,1]. Many sources relax the boundary to even [-3,3]. The PRES variable is minmax scaled to bound the transformed variable within [0,1]. Minmax scaling of a variable *x* is done as follows:

$$x_{scaled} = \frac{x - x_{min}}{x_{max} - x_{min}}$$

The preceding formula would result in $\texttt{x\_scaled} \in [0,1]$:

```
from sklearn.preprocessing import MinMaxScaler
scaler = MinMaxScaler(feature_range=(0, 1))
df['scaled_PRES'] = scaler.fit_transform(np.array(df['PRES'])
  .reshape(-1, 1))
```

The variable `scaled_PRES` is used for developing the time series forecasting model. For making prediction on the original variable `PRES`, we apply suitable inverse transformation on the model's prediction.

Before training the model, the dataset is split into two parts, train set and validation set. The neural network is trained on the train set. This means that computation of the loss function, backpropagation, and weight updates by a gradient descent algorithm is done on the train set. The validation set is used to evaluate the model and to determine the number of epochs of training. Increasing the number of epochs will further decrease the loss function on the train set but might not necessarily have the same effect for the validation set due to overfitting on the train set. Hence, the number of epochs is controlled by keeping a tap on the loss function computed for the validation set. We use Keras with TensorFlow backend to define and train the model. All the steps involved in model training and validation is done by calling appropriate functions of the Keras API.

Let's start by splitting the data into train and validation sets. The dataset's time period is from

Jan 1st, 2010 to Dec 31st, 2014. The first four years-2010 to 2013-is used as train and

2014 is kept for validation:

```
split_date = datetime.datetime(year=2014, month=1, day=1, hour=0)
df_train = df.loc[df['datetime']<split_date]
df_val = df.loc[df['datetime']>=split_date]
print('Shape of train:', df_train.shape)
print('Shape of test:', df_val.shape)
```

The preceding lines of code generate the following output:

```
Shape of train: (35064, 15)
Shape of test: (8760, 15)
```

The train and validation time series of `scaled_PRES` are also plotted:

```
plt.figure(figsize=(5.5, 5.5))
g = sns.tsplot(df_train['scaled_PRES'], color='b')
g.set_title('Time series of scaled Air Pressure in train set')
g.set_xlabel('Index')
g.set_ylabel('Scaled Air Pressure readings')
```

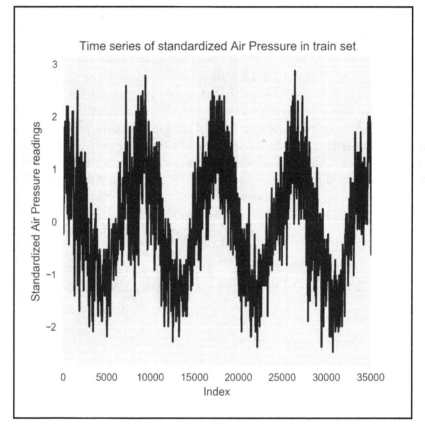

Figure 5.8: Time series of minmax scaled air pressure in training data

```
plt.figure(figsize=(5.5, 5.5))
g = sns.tsplot(df_val['scaled_PRES'], color='r')
g.set_title('Time series of standardized Air Pressure in validation set')
g.set_xlabel('Index')
g.set_ylabel('Standardized Air Pressure readings')
```

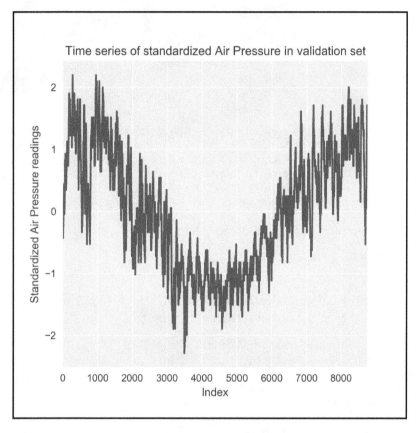

Figure 5.9: Time series of minmax scaled air pressure in validation data

Now we need to generate regressors (X) and target variable (y) for train and validation. A 2D array of regressors and 1-D array of target is created from the original 1-D array of the scaled_PRES column in the DataFrames. For the time series forecasting model, the past seven days of observations are used to predict for the next day. This is equivalent to an AR(7) model. We define a function that takes the original time series and the number of timesteps as input to return the arrays of X and y:

```
def makeXy(ts, nb_timesteps):
    """
    Input:
```

```
        ts: original time series
        nb_timesteps: number of time steps in the regressors
    Output:
        X: 2-D array of regressors
        y: 1-D array of target
    """
    X = []
    y = []
    for i in range(nb_timesteps, ts.shape[0]):
        X.append(list(ts.loc[i-nb_timesteps:i-1]))
        y.append(ts.loc[i])
    X, y = np.array(X), np.array(y)
    return X, y
```

The makeXy function is called to generate the train and validation sets:

```
X_train, y_train = makeXy(df_train['scaled_PRES'], 7)
print('Shape of train arrays:', X_train.shape, y_train.shape)
X_val, y_val = makeXy(df_val['scaled_PRES'], 7)
print('Shape of validation arrays:', X_val.shape, y_val.shape)
```

The output of the preceding print function is as follows:

```
Shape of train arrays: (35057, 7) (35057,)
Shape of validation arrays: (8753, 7) (8753,)
```

Now we define the MLP using the Keras functional API. In this approach, layers are declared and cascaded as input and output of each other:

```
from keras.layers import Dense, Input, Dropout
from keras.optimizers import SGD
from keras.models import Model
from keras.models import load_model
from keras.callbacks import ModelCheckpoint
```

The input layer is declared with shape (None, 7) and of type float32. None indicates the number of instances that is determined at runtime:

```
input_layer = Input(shape=(7,), dtype='float32')
```

Dense layers are declared with linear activation:

```
dense1 = Dense(32, activation='linear')(input_layer)
dense2 = Dense(16, activation='linear')(dense1)
dense3 = Dense(16, activation='linear')(dense2)
```

Multiple hidden layers and large number of neurons in each hidden layer gives neural networks the ability to model complex non-linearity of the underlying relations between regressors and the target. However, deep neural networks can also overfit train data and give poor results on validation or test set. Dropout has been used to regularize deep neural networks. In this example, a dropout layer is added before the output layer. Dropout randomly sets p fraction of input neurons to zero before passing to the next layer. Randomly dropping inputs essentially acts as bootstrap aggregating or bagging type of model ensembling. Random forest uses bagging by building trees on random subsets of input features. We use p=0.2 to dropout 20% of randomly selected input features:

```
dropout_layer = Dropout(0.2)(dense3)
```

Finally, the output layer gives prediction for the next day's air pressure:

```
output_layer = Dense(1, activation='linear')(dropout_layer)
```

The input, dense, and output layers will now be packed inside a `Model`, which is the wrapper class to train and make predictions. **Mean square error** (**MSE**) is used as the `loss` function.

The network's weights are optimized by the **Adam algorithm**. Adam stands for adaptive moment estimation and has been a popular choice to train deep neural networks. Unlike stochastic gradient descent, Adam uses different learning rates for each weight and separately updates them as the training progresses. The learning rate of a weight is updated based on exponentially weighted moving averages of the weight's gradients and the squared gradients:

```
ts_model = Model(inputs=input_layer, outputs=output_layer)
ts_model.compile(loss='mean_squared_error', optimizer='adam')
ts_model.summary()
```

The `summary` function displays layer-wise details such as the shape of input and output and number of trainable weights:

Layer (type)	Output Shape	Param #
input_6 (InputLayer)	(None, 7)	0
dense_21 (Dense)	(None, 32)	256
dense_22 (Dense)	(None, 16)	528
dense_23 (Dense)	(None, 16)	272
dropout_6 (Dropout)	(None, 16)	0

```
dense_24 (Dense)              (None, 1)                    17
=================================================================
Total params: 1,073
Trainable params: 1,073
Non-trainable params: 0
```

The model is trained by calling the `fit` function on the model object and passing the **X_train** and **y_train**. The training is done for a predefined number of epochs. Additionally, **batch_size** defines the number of samples of train set to be used for an instance of backpropagation. The validation dataset is also passed to evaluate the model after every epoch completes. A **ModelCheckpoint** object tracks the loss function on the validation set and saves the model for the epoch at which the loss function has been minimum:

```
save_weights_at = os.path.join('keras_models',
'PRSA_data_Air_Pressure_MLP_weights.{epoch:02d}-{val_loss:.4f}.hdf5')
save_best = ModelCheckpoint(save_weights_at, monitor='val_loss', verbose=0,
                           save_best_only=True, save_weights_only=False,
mode='min',
                           period=1)
ts_model.fit(x=X_train, y=y_train, batch_size=16, epochs=20,
             verbose=1, callbacks=[save_best], validation_data=(X_val,
y_val),
             shuffle=True)
```

The epoch-wise MSE on the train and validation sets are detailed in the Jupyter notebook, `code/ Chapter_5_Air Pressure_Time_Series_Forecasting_by_MLP.ipynb`. The time taken to complete each epoch is also shown.

Predictions are made from the best saved model. The model's predictions, which are on the standardized air pressure, are inversely transformed to get predictions on the original air pressure. The goodness-of-fit or R-squared is also calculated:

```
best_model = load_model(os.path.join('keras_models',
'PRSA_data_Air_Pressure_MLP_weights.06-0.0039.hdf5'))
preds = best_model.predict(X_val)
pred_PRES = mu + sigma*preds
pred_PRES = np.squeeze(pred_PRES)

from sklearn.metrics import r2_score
r2 = r2_score(df_val['PRES'].loc[7:], pred_PRES)
print('R-squared for the validation set:', round(r2,4))
```

The R-squared of the model on the validation set is 0.9957. Lastly, the first fifty actual and predicted values of air pressure are plotted:

```
plt.figure(figsize=(5.5, 5.5))
plt.plot(range(50), df_val['PRES'].loc[7:56], linestyle='-', marker='*',
color='r')
plt.plot(range(50), pred_PRES[:50], linestyle='-', marker='.', color='b')
plt.legend(['Actual','Predicted'], loc=2)
plt.title('Actual vs Predicted Air Pressure')
plt.ylabel('Air Pressure')
plt.xlabel('Index')
```

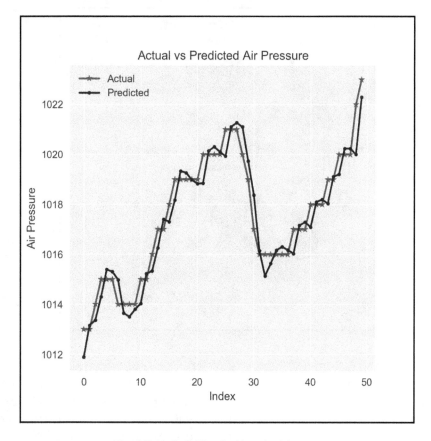

Figure 5.10: Actual and MLP-predicted time series of air pressure

A time series forecasting model using MLP is also trained for the pm2.5 variable and the detailed implementation is in the Jupyter notebook, `code\Chapter_5_PM2.5_Time_Series_Forecasting_by_MLP.ipynb`.

A boxplot for pm2.5 is plotted to check the presence of outliers:

```
plt.figure(figsize=(5.5, 5.5))
g = sns.boxplot(df['pm2.5'])
g.set_title('Box plot of pm2.5')
```

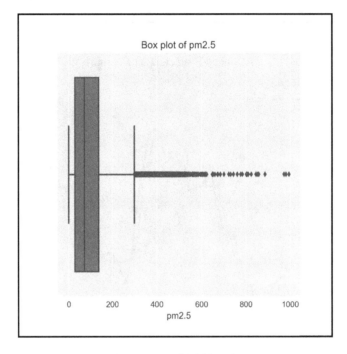

Figure 5.11: Boxplot of PM2.5 time series

As seen in the preceding figure, pm2.5 has outliers and therefore, the choice of MSE as a loss function to train the MLP is not the best. MSE, being a square of the deviations between actual and predicted, gives huge fluctuations to the loss function. This destabilizes the gradient descent algorithm and adversely affects its convergence. MAE absolute being first order difference, is less susceptible to fluctuations due to outliers. Therefore, in this case, MAE is used to train the neural network.

Before proceeding to defining and training MLPs, let's have a close look at the time series for over a year and for over six months to see if any patterns, such as trend, seasonality, and so on, apparently exist:

```
plt.figure(figsize=(5.5, 5.5))
g = sns.tsplot(df['pm2.5'].loc[df['datetime']<=datetime.datetime(year=2010,
month=6,day=30)], color='g')
g.set_title('pm2.5 during 2010')
g.set_xlabel('Index')
g.set_ylabel('pm2.5 readings')

plt.figure(figsize=(5.5, 5.5))
g = sns.tsplot(df['pm2.5'].loc[df['datetime']<=datetime.datetime(year=2010,
month=1,day=31)], color='g')
g.set_title('pm2.5 during Jan 2010')
g.set_xlabel('Index')
g.set_ylabel('pm2.5 readings')
```

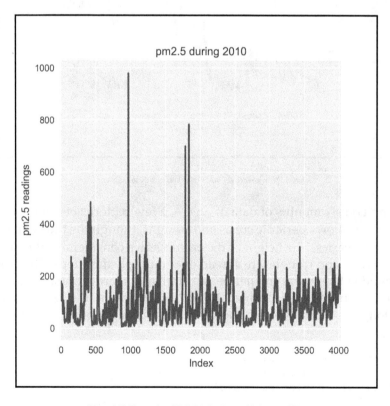

Figure 5.12: Time series of PM2.5 during January 2010 to June 2010

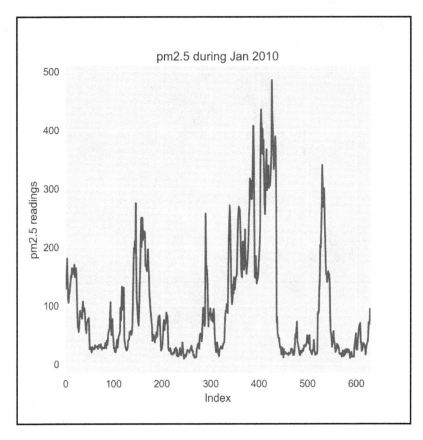

Figure 5.13: Time series of PM2.5 during January 2010 to December 2010

When we zoom in on six months of data of pm2.5, a few patterns become apparent. The time series of pm2.5 shows periodic crests and troughs, though the time gap between two highs and two lows varies. The height of the crests shows considerable fluctuations. Moreover, both crests and troughs are spread over multiple timesteps. This is indicated by a tall peak surrounded by several smaller peaks. Similarly, there are small fluctuations in a trough as well. The entire series, during 2010 to 2014, as seen in the following figure, does not show any long-term trend though short-term trends might be present throughout the series:

```
plt.figure(figsize=(5.5, 5.5))
g = sns.tsplot(df['pm2.5'])
g.set_title('Time series of pm2.5')
g.set_xlabel('Index')
g.set_ylabel('pm2.5 readings')
```

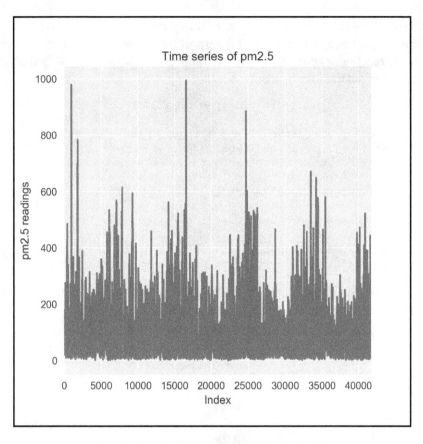

Figure 5.14: Time series of PM2.5 during January 2010 to December 2014

A neural network with multiple hidden layers and multiple neurons in each hidden layer would be suitable for the modeling of such a complex nonlinear pattern in the data. Let's give it a try using MLP.

The MLP of **pm2.5** has three dense layers with thirty-two, sixteen, and sixteen hidden neurons in the first, second, and third layer, respectively. Each layer has `tanh` activation. The output layer has linear activation. The neural network is trained using the Adam optimizer with MAE as the loss function:

```
input_layer = Input(shape=(7,), dtype='float32')

dense1 = Dense(32, activation='tanh')(input_layer)
dense2 = Dense(16, activation='tanh')(dense1)
dense3 = Dense(16, activation='tanh')(dense2)

dropout_layer = Dropout(0.2)(dense3)

output_layer = Dense(1, activation='linear')(dropout_layer)

ts_model = Model(inputs=input_layer, outputs=output_layer)
ts_model.compile(loss='mean_absolute_error', optimizer='adam')
```

The best model gives MAE of 11.8993 on the validation set. The first fifty actual and predicted values are plotted in the following figure:

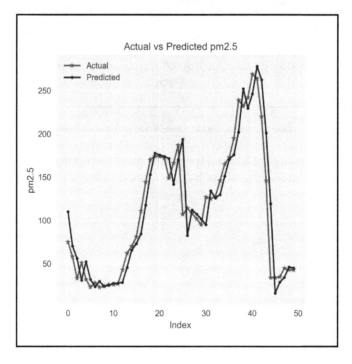

Figure 5.15: Actual and MLP-predicted time series of PM2.5

Recurrent neural networks

So far, we have used an MLP to develop a time series forecasting model. To predict the series P at time $[x_{t-1}, \ldots, x_{t-p}]$ we fed an input vector of past P time steps P to an MLP. The past P time steps are fed to the MLP as uncorrelated independent variables. One problem with this kind of model is that it does not implicitly consider the sequential nature of the time series data where observations have correlation with each other. The correlation in a time series can also be interpreted as the memory that the series carries over itself. In this section, we will discuss **recurrent neural networks (RNNs)** that are architecturally different from MLPs and are more appropriate to fit sequential data.

RNNs emerged as a good choice to develop language models that model the probability of occurrence of a word given the words that appear prior to it. So far, RNNs have been used to develop models that do text classification, for example, sentiment prediction, language translation, and, in conjunction with convolutional neural networks, to describe images through text generation.

The following figure shows an RNN having $\hat{x}_t$ timesteps each of which is fed with the input occurring at the corresponding step. The output from the last timestep is the prediction from the input series. For example, this RNN can be used to develop a time series forecasting model where the input series $[x_{t-1}, \ldots, x_{t-p}]$ is fed to the RNN and the output from the last timestep is the prediction :

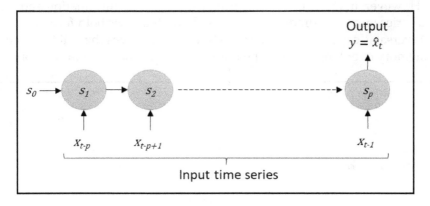

Figure 5.16: Recurrent neural network with p timesteps

The internal state of each computational unit of an RNN is s_0 and carries the memory of the series. The number of hidden neurons in each timestep is the dimension of the internal state, which is calculated as f. This internal state of each timestep carries the memory of the series. For the first timestep, the input $y = h(Vs_{t-1})$ is initialized to all zeros. The function $h(\cdot)$ is a nonlinear transformation such as `tanh`. The output of the last timestep is a function of the last internal state U, where W is a nonlinear transformation. The RNN can also be made to return output from every timestep and this is useful for language translations models that require the translated text to be a series of words in the target language. An important thing to note is that the weights V, x_i, and x_i are shared across the timesteps of the RNN and this way the number of trainable parameters of the network is kept lower.

In case of language models, the input m could be a one-hot encoded representation of the words. For time series modeling of one variable, x_i is a single number. However, RNNs can be applied to multivariate time series and this is particularly useful to address cross-correlation between the individual sequences. If the RNN models m time series, then x_i is an m-dimensional vector.

Bi-directional recurrent neural networks

The RNN discussed so far is unidirectional and traverses along the direction of the original time series. However, in many cases, capturing the sequential information and memory in the reversed series improves prediction. Such an RNN that uses both forward and backward traversal is called a bi-directional RNN that improves the ability of the network to capture memory over long ranges. The following figure shows a bi-directional RNN:

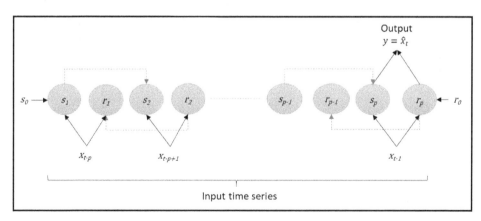

Figure 5.17: Bi-directional recurrent neural network with *p* timesteps

Deep recurrent neural networks

The power of deep learning comes with stacking multiple computational layers on top of each other. In case of MLPs, multiple hidden layers are placed against each other. We can make deep RNNs by stacking multiple RNNs on top of each other. In a deep RNN, the input sequence for a recurrent layer is the output sequence of the previous recurrent layer. The final prediction is taken from the last timestep of the final RNN layer. The following figure illustrates a deep RNN:

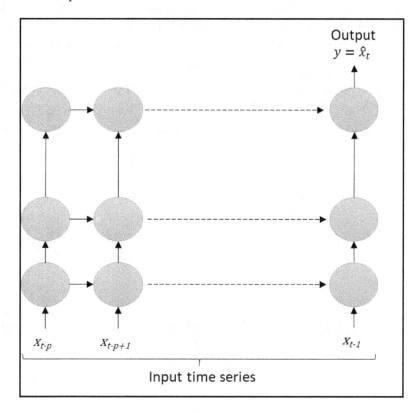

Figure 5.18: Deep recurrent neural network with *p* timesteps

We can create deep bi-directional RNN as well by stacking multiple bi-directional RNNs on top of each other. Such deep RNNs are used for complex tasks such as language translation and text generation to describe a context that can be an image, for example; quite evidently, deep RNNs also require millions of network weights to be trained.

Training recurrent neural networks

RNNs are notoriously difficult to be trained. Vanilla RNNs, the kind of RNNs we have been talking about so far, suffer from vanishing and exploding gradients that give erratic results during training. As a result, RNNs have difficulty in learning long-range dependencies. For time series forecasting, going too many timesteps back in the past would be problematic. To address this problem, **Long Short Term Memory (LSTM)** and **Gated Recurrent Unit (GRU)**, which are special types of RNNs, have been introduced. In this chapter, we will use LSTM and GRU to develop the time series forecasting models. Before this, let's review how RNNs are trained using **Backpropagation Through Time (BPTT)**, a variant of the backpropagation algorithm. We will find out how vanishing and exploding gradients arise during BPTT.

Let's consider the computational graph of the RNN that we have been using for the time series forecasting. The gradient computation is shown in the following figure:

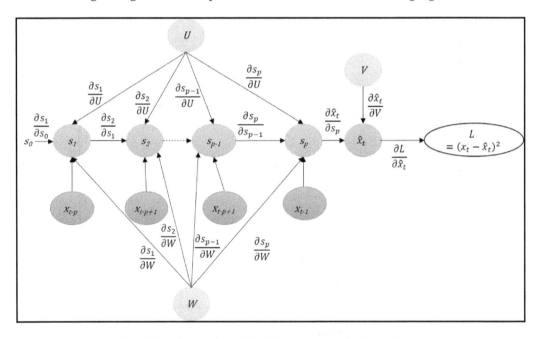

Figure 5.19: Back-Propagation Through Time for deep recurrent neural network with p timesteps

For weights U, there is one path to compute the partial derivative, which is given as follows:

$$\frac{\partial L}{\partial V} = \frac{\partial L}{\partial \hat{x}_t} \frac{\partial \hat{x}_t}{\partial V}$$

However, due to the sequential structure of the RNN, there are multiple paths connecting the weights and the loss and and the loss. Hence, the partial derivative is the sum of partial derivatives along the individual paths that starts at the loss node and ends at every timestep node in the computational graph:

$$\frac{\partial L}{\partial W} = \left(\frac{\partial L}{\partial \hat{x}_t}\frac{\partial \hat{x}_t}{\partial s_p}\frac{\partial s_p}{\partial W}\right) + \left(\frac{\partial L}{\partial \hat{x}_t}\frac{\partial \hat{x}_t}{\partial s_p}\frac{\partial s_p}{\partial s_{p-1}}\frac{\partial s_{p-1}}{\partial W}\right) + \cdots + \left(\frac{\partial L}{\partial \hat{x}_t}\frac{\partial \hat{x}_t}{\partial s_p}\frac{\partial s_p}{\partial s_{p-1}}\cdots\frac{\partial s_2}{\partial s_1}\frac{\partial s_1}{\partial W}\right)$$

$$\frac{\partial L}{\partial U} = \left(\frac{\partial L}{\partial \hat{x}_t}\frac{\partial \hat{x}_t}{\partial s_p}\frac{\partial s_p}{\partial U}\right) + \left(\frac{\partial L}{\partial \hat{x}_t}\frac{\partial \hat{x}_t}{\partial s_p}\frac{\partial s_p}{\partial s_{p-1}}\frac{\partial s_{p-1}}{\partial U}\right) + \cdots + \left(\frac{\partial L}{\partial \hat{x}_t}\frac{\partial \hat{x}_t}{\partial s_p}\frac{\partial s_p}{\partial s_{p-1}}\cdots\frac{\partial s_2}{\partial s_1}\frac{\partial s_1}{\partial U}\right)$$

The technique of computing the gradients for the weights by summing over paths connecting the loss node and every timestep node is backpropagation through time,which is a special case of the original *backpropagation* algorithm. The problem of vanishing gradients in long-range RNNs is due to the multiplicative terms in the BPTT gradient computations. Now let's examine a multiplicative term from one of the preceding equations.

The gradients along the computation path connecting the loss node and i^{th} timestep is $\frac{\partial L}{\partial \hat{x}_t}\frac{\partial \hat{x}_t}{\partial s_p}\frac{\partial s_p}{\partial s_{p-1}}\cdots\frac{\partial s_{i+1}}{\partial s_i}\frac{\partial s_i}{\partial s_{i-1}}\cdots\frac{\partial s_i}{\partial W}$. This chain of gradient multiplication is ominously long to model long-range dependencies and this is where the problem of vanishing gradient arises.

The activation function for the internal state $[0,1)$ is either a tanh or sigmoid. The first derivative of tanh is $(\frac{1}{1+e^{-x}})^2 e^{-x}$, which is bound in $(0,\frac{1}{4}]$. For the sigmoid function, the first-order derivative is $\frac{\partial s_i}{\partial s_{i-1}}$, which is bound in s_i. Hence, the gradients W are positive fractions. For long-range timesteps, multiplying these fractional gradients diminishes the final product to zero and there is no gradient flow from a long-range timestep. Due to the negligibly low values of the gradients, the weights do not update and hence the neurons are said to be saturated.

It is noteworthy that U, $\frac{\partial s_i}{\partial s_{i-1}}$, and $\frac{\partial s_i}{\partial W}$ are matrices and therefore the partial derivatives t and s_t are computed on matrices. The final output is computed through matrix multiplications and additions. The first derivative on matrices is called **Jacobian**. If any one element of the Jacobian matrix is a fraction, then for a long-range RNN, we would see a vanishing gradient. On the other hand, if an element of the Jacobian is greater than one, the training process suffers from exploding gradients.

Solving the long-range dependency problem

We have seen in the previous section that it is difficult for vanilla RNNs to effectively learn long-range dependencies due to vanishing and exploding gradients. To address this issue, Long Short Term Memory network was developed in 1997 by Sepp Hochreiter and Jürgen Schmidhuber. Gated Recurrent Unit was introduced in 2014 and gives a simpler version of LSTM. Let's review how LSTM and GRU solve the problem of learning long-range dependencies.

Long Short Term Memory

LSTM introduces additional computations in each timestep. However, it can still be treated as a black box unit that, for timestep (h_t), returns the state internal(f_t) and this is forwarded to the next timestep. However, internally, these vectors are computed differently. LSTM introduces three new gates: the input (o_t), forget (g_t), and output (c_t) gates. Every timestep also has an internal hidden state (h_t) and internal memory $(\sigma(W^h x_t + U^h s_{t-1})$. These new units are computed as follows:

$$f_t = \sigma(W^f x_t + U^f s_{t-1})$$

$$o = \sigma(W^o x_t + U^o s_{t-1})$$

$$g_t = tanh(W^g x_t + U^g s_{t-1})$$

$$c_t = f_t \odot c_{t-1} + h_t \odot g_t$$

$$s_t = tanh(c_t) \odot o_t$$

$$h_t = f_t$$

Now, let's understand the computations. The gates o_t, $\sigma(\cdot)$, and $[0,1]$ are generated through sigmoid activations h_t that limits their values within f_t. Hence, these act as gates by letting out only a fraction of the value when multiplied with another variable. The input gate o_t controls the fraction of the newly computed input to keep. The forget gate c_t determines the effect of the previous timestep and the output gate f_t controls how much of the internal state to let out. The internal hidden state is calculated from the input to the current timestep and output of the previous timestep. Note that this is the same as computing the internal state in a vanilla RNN. h_t is the internal memory unit of the current timestep and considers the memory from the previous step but downsized by a fraction s_t and the effect of the internal hidden state but mixed with the input gate o_t. Finally, z_t would be passed to the next timestep and computed from the current internal memory and the output gate r_t. The input, forget, and output gates are used to selectively include the previous memory and the current hidden state that is computed in the same manner as in vanilla RNNs. The gating mechanism of LSTM allows memory transfer from over long-range timesteps.

Gated Recurrent Units

GRU is simpler than LSTM and has only two internal gates, namely, the update gate (z_t) and the reset gate (r_t). The computations of the update and reset gates are as follows:

$$z_t = \sigma(W^z x_t + U^z s_{t-1})$$

$$r_t = \sigma(W x_t + U s_{t-1})$$

The state s_t of the timestep t is computed using the input x_t , state s_{t-1} from the previous timestep, the update, and the reset gates:

$$s_t = z_t \circ s_{t-1} + (1 - z_t) \circ \tanh\left((W^h x_t + U^h (r_t \circ s_{t-1})\right)$$

The update being computed by a sigmoid function determines how much of the previous step's memory is to be retained in the current timestep. The reset gate controls how to combine the previous memory with the current step's input.

Compared to LSTM, which has three gates, GRU has two. It does not have the output gate and the internal memory, both of which are present in LSTM. The update gate in GRU determines how to combine the previous memory with the current memory and combines the functionality achieved by the input and forget gates of LSTM. The reset gate, which combines the effect of the previous memory and the current input, is directly applied to the previous memory. Despite a few differences in how memory is transmitted along the sequence, the gating mechanisms in both LSTM and GRU are meant to learn long-range dependencies in data.

Which one to use - LSTM or GRU?

Both LSTM and GRU are capable of handling memory over long RNNs. However, a common question is which one to use? LSTM has been long preferred as the first choice for language models, which is evident from their extensive use in language translation, text generation, and sentiment classification. GRU has the distinct advantage of fewer trainable weights as compared to LSTM. It has been applied to tasks where LSTM has previously dominated. However, empirical studies show that neither approach outperforms the other in all tasks. Tuning model hyperparameters such as the dimensionality of the hidden units improves the predictions of both. A common rule of thumb is to use GRU in cases having less training data as it requires less number of trainable weights. LSTM proves to be effective in case of large datasets such as the ones used to develop language translation models.

Recurrent neural networks for time series forecasting

We will continue to use the dataset on air pollution to demonstrate recurrent neural networks for time series forecasting. LSTM is used to forecast air pressure and GRU is demonstrated on pm2.5.

Reading and preprocessing the data remains the same as we have done for the examples on MLPs. The original dataset is split into two sets-train and validation, which are used for model training and validation respectively.

The `makeXy` function is used to generate arrays of regressors and targets-`X_train`, `X_val`, `y_train` and `y_val`. `X_train`, and `X_val`, as generated by the `makeXy` function, are 2D arrays of shape `(number of samples, number of timesteps)`. However, the input to RNN layers must be of shape `(number of samples, number of timesteps, number of features per timestep)`. In this case, we are dealing with only `pm2.5`, hence `number of features per timestep` is one. `Number of timesteps` is seven and `number of samples` is the same as the `number of samples` in `X_train` and `X_val`, which are reshaped to 3D arrays:

```
X_train, X_val = X_train.reshape((X_train.shape[0], X_train.shape[1], 1)),
                 X_val.reshape((X_val.shape[0], X_val.shape[1], 1))
print('Shape of 3D arrays:', X_train.shape, X_val.shape)
```

`X_train` and `X_val` have been reshaped to 3D arrays and their new shapes are seen in the output of the preceding `print` statement, which is as follows:

```
Shape of 3D arrays: (35057, 7, 1) (8753, 7, 1)
```

To add an LSTM layer to the neural network, we need to import the LSTM class from Keras:

```
from keras.layers.recurrent import LSTM
```

The neural network to develop the time series forecasting model has an input layer, which feeds into the LSTM layer. The LSTM layer has seven timesteps, which is the same as the number of historical observations taken to make the next-day prediction of air pressure. Only the last timestep of the LSTM returns an output. There are sixty-four hidden neurons in each timestep of the LSTM layer. Hence, the output from the LSTM has sixty-four features:

```
input_layer = Input(shape=(7,1), dtype='float32')
lstm_layer = LSTM(64, input_shape=(7,1),
return_sequences=False)(input_layer)
```

Next, the LSTM's output is passed to a dropout layer that randomly drops 20% of the input before passing to the output layer, which has a single hidden neuron with a linear activation function:

```
dropout_layer = Dropout(0.2)(lstm_layer)
output_layer = Dense(1, activation='linear')(dropout_layer)
```

Finally, all the layers are wrapped in a `keras.models.Model` and trained for twenty epochs to minimize MSE using the Adam optimizer:

```
ts_model = Model(inputs=input_layer, outputs=output_layer)
ts_model.compile(loss='mean_squared_error', optimizer='adam')
```

We have used `keras.callbacks.ModelCheckpoint` as a callback to track the MSE of the model on the validation set and save weights from the epoch that gives minimum validation error. R-squared of the best model for the validation set is 0.9959. The first fifty actual and predicted values are shown in the following figure:

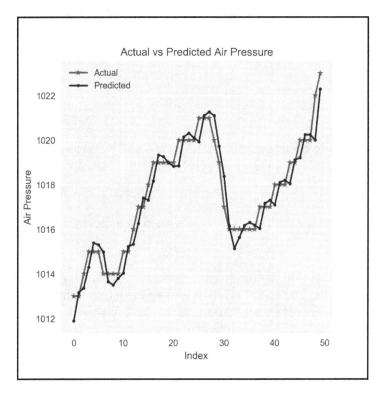

Figure 5.20: Actual and LSTM-predicted time series of air pressure

 The code for this example can be found in the Jupyter notebook, `code/Chapter_5_Air Pressure_Time_Series_Forecasting_by_LSTM.ipynb`.

We have used two stacked GRU layers to develop an RNN-based time series forecasting model of `pm2.5`:

```
gru_layer1 = GRU(64, input_shape=(7,1), return_sequences=True)(input_layer)
gru_layer2 = GRU(32, input_shape=(7,64),
return_sequences=False)(gru_layer1)
```

The first GRU takes a sequential input from the preceding input layer. Each timestep of the first GRU returns a sixty-four-dimensional feature vector as output. This sequence is passed as input to the next GRU layer. The second GRU layer returns output only from the last timestep. The neural network is trained to minimize the MAE loss using the Adam optimizer.

The best MAE obtained for the validation set is 11.388.

The full code for this example is in the Jupyter notebook,
`code/Chapter_5_PM2.5_Time_Series_Forecasting_by_GRU.ipynb`.

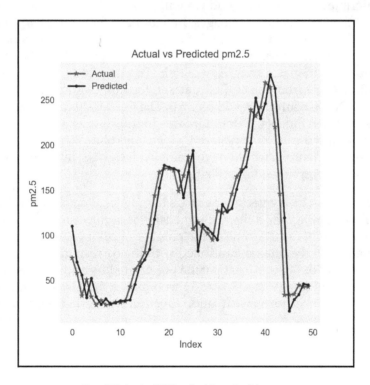

Figure 5.21: Actual and GRU-predicted time series of air pressure

Convolutional neural networks

This section describes **Convolutional Neural Networks** (**CNNs**) that are primarily applied to develop supervised and unsupervised models when the input data are images. In general, **two-dimensional** (**2D**) convolutions are applied to images but **one-dimensional** (**1D**) convolutions can be used on a sequential input to capture time dependencies. This approach is explored in this section to develop time series forecasting models.

2D convolutions

Let's start by describing the 2D CNNs and we will derive 1D CNNs as a special case. CNNs take advantage of the 2D structure of images. Images have a rectangular dimension of w, where n is the height and $h \times w \times n$ is the width of the image. The color value of every pixel would be an input feature to the model. Using a fully-connected dense layer having 28 x 28 neurons, the number of trainable weights would be 28 x 28 x 100 = 78400. For images of handwritten digits 32 x 32 from the MNIST dataset, the number of trainable weights in the first dense layer with 100 neurons would be $c = 3$. The CIFAR-10 dataset is popularly used to train object recognition models. Colored images in this dataset are 32 x 32 x 3 x 100 = 307200 and have $m \times m \times 3$ color channels-red, green, and blue. Hence, a fully-connected dense layer having 100 neurons would have trainable weights. Hence, training dense layers for images becomes a computational challenge.

CNNs solve this problem by connecting neurons to only local patches of the image. As shown in the following diagram, a filter of dimension is applied to a local image patch. The third dimension of the filter is the same as the number of color channels of the image. The weight of each neuron in the filter is multiplied with the corresponding pixel value in the image. The final feature out of this local patch is computed by adding these individual values and optionally adding with a bias and then passing the sum through an activation function. The popular choice for the activation function is **rectified linear units** (**ReLu**):

$ReLu(z) = 0 \; if \; z \leq 0$

$= z \; if \; z > 0$

The first-order derivative of ReLu is either 0 or 1 and has good gradient flow properties. Hence, it is preferred in training deep convolutional networks having multiple layers. The filters as shown in the following figure are a convolutional layer. To learn different features, convolutional layers generally have multiple filters.

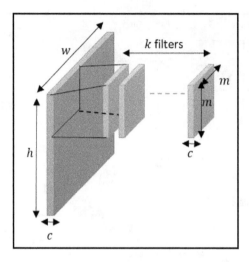

Figure 5.22: Application of filters for feature generation by convolutional neural layers

To cover the entire image, the filter is moved with a horizontal stride of *4 x 4* pixel units and vertical stride of *2 x 2* pixel units. The movement of the filter over the image is called convolution. The features resulting from convolutions over the entire image forms a rectangular feature map. Now let's understand how a convolution generates the feature map. For this, let's consider a image with a one-color channel and having pixel values as shown in the following figure. We apply a *3 x 3* convolution filter and move it horizontally by one pixel unit and vertically by one pixel unit.

The first convolution is calculated as *1×-1+2×1+2×2+1×1=6*. The features from other convolutions are also calculated in a similar way. In this example, all calculated values being positive, the ReLu activation acts like an identity function. This convolution process creates a feature map, as shown in the following figure:

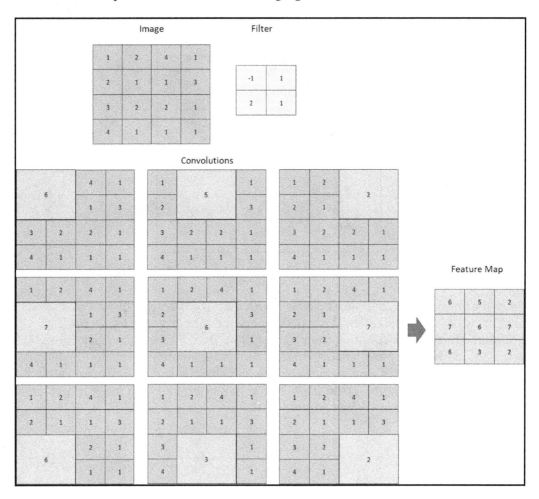

Figure 2.23: Illustration of 2-D convolution

Note that a *2 x 2* filter with unit pixel horizontal and vertical strides shrinks the size of the feature map to *3 x 3* from the original image that was *6 x 6*. To ensure that the feature map is of the same size as that of the image, we could have concatenated zero-valued pixels across the horizontal and vertical borders of the image and used a filter. Adding zero-valued pixels is called zero-padding that, in this case, would have made the input image of a dimension of *2 x 2*. The feature map generated from the zero-padded image is shown in the following figure:

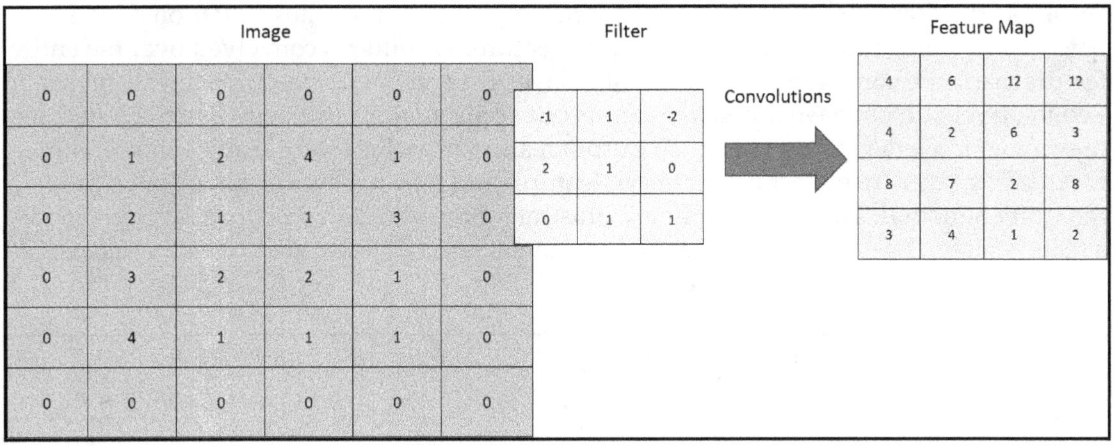

Figure 2.24: Feature map generated by 2-D convolution

The feature map, generated by a convolution layer, would be fed to a downstream convolutional layer just as the original input has been given to the first convolution layer. Multiple convolution layers, stacked against each other, generate better features from the original images. These features are then passed to downstream fully-connected dense layers that generate a softmax output over the set of object classes.

In most deep learning models on images, the convolutional layer is not used to downsample the original image or the intermediate feature maps. If this is the case, when output from convolutional layers is fed to dense layers, the number of trainable weights would still be huge. Then what does the convolutional layers achieve? How do we downsample feature maps to reduce the number of trainable weights in the dense layers?

Firstly, though we can achieve downsampling in the convolutional layers, these are preferably used to extract image features such as edges, corners, shapes, and so on. Secondly, the downsampling achieved by a pooling layer that applies a filter on a local patch of a feature map and computes a single feature. The filter is convolved over the entire feature map. The convolution by the pooling layer generates a downsampled feature map. Pooling layers do not have trainable weights but apply simple arithmetic functions such as maximum or average to generate their output feature map. Empirical studies show maxpooling layers to extract useful image features and lead to better accuracy of image recognition models. The following figure illustrates how a *2 x 2* maxpooling layer generates a feature map. The maxpooling layer has horizontal and vertical strides of two units and has downsampled the input feature map from to *28 x 28*.

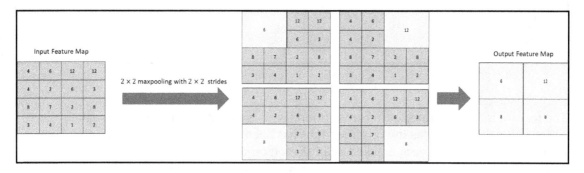

Figure 2.25: Illustration of 2-D maxpooling

The following figure illustrates a neural network for image recognition. There are two blocks of convolutional layers followed by a block of fully-connected dense layers. Labeled images are fed to this network and the output is probabilities over the label classes. For example, a digit recognition model can be trained using the MNIST dataset that has *t* grayscale images of handwritten digits (0, 1, 2, ... 9). The grayscale images are fed to a neural network, as shown in the following figure. The last layer of the network is a softmax layer that gives predicted probabilities over the ten-digit classes. The class with the highest probability is the predicted digit.

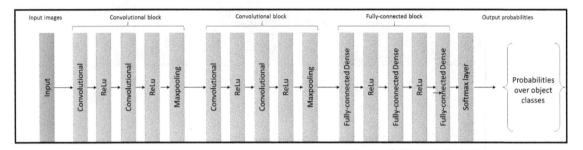

Figure 2.26: Architecture of deep convolutional neural network for image classification

1D convolution

1D convolution layers can be used to develop time series forecasting models. A time series having 1 x *m* observations is like an image of dimension *p*, which has a height of a single pixel. In this case, 1D convolution can be applied as a special case of 2D convolution using a *1 x 3* filter. Additionally, the filter is moved only along the horizontal direction by strides length of *1 x 8* time units.

Let's understand how 1D convolution works. Consider the following figure that shows a time series of ten timesteps. A $(1 \times -1) + (2 \times 1) + (-1 \times 2) = -1$ filter is moved by a stride of one time unit over the series. Thus, a *1 x 3* feature map is generated. The first element of the feature map is computed as *1 x 10*. The rest of the timesteps are computed in a similar way. We have not zero-padded the original time series so the feature map is two units shorter than the original series. However, zero-padding the start and end of the input time series and using the same filter would have resulted in a *1 x 3* feature map. A pooling layer can be stacked with the convolutional layer to downsample the feature map.

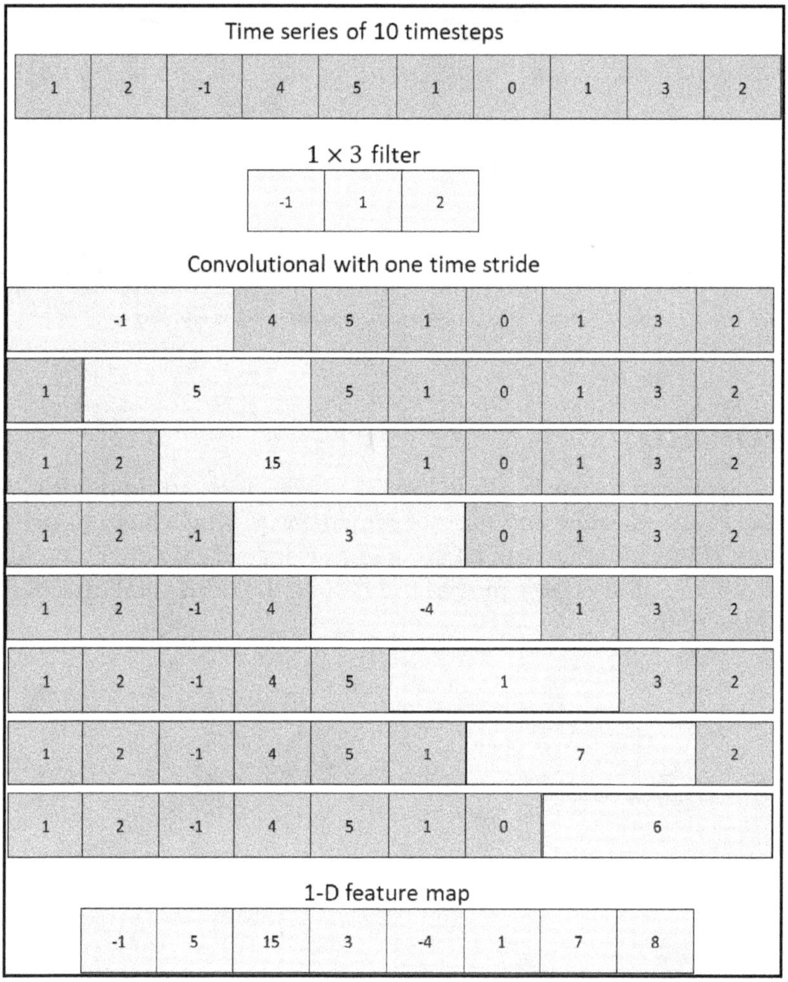

Time series of 10 timesteps

1	2	-1	4	5	1	0	1	3	2

1 × 3 filter

-1	1	2

Convolutional with one time stride

-1			4	5	1	0	1	3	2
1		5		5	1	0	1	3	2
1	2		15		1	0	1	3	2
1	2	-1		3		0	1	3	2
1	2	-1	4		-4		1	3	2
1	2	-1	4	5		1		3	2
1	2	-1	4	5	1		7		2
1	2	-1	4	5	1	0		6	

1-D feature map

-1	5	15	3	-4	1	7	8

Figure 2.27: Illustration of 1-D convolution

The approach of using a *1 x 3* convolution filter is equivalent to training several local auto-regressive models of order three. These local models generate features over short-term subsets of the input time series. When an average pooling layer is used after the 1D convolution layer, it creates moving averages over the feature map generated by the preceding convolution layer. Furthermore, several 1D convolution and pooling layers, when stacked with each other, give a powerful way of extracting features from the original time series. Thus, using CNNs proves to be effective when dealing with complex, nonlinear time series such as audio waves, speech, and so on. As a matter of fact, CNNs have been successfully applied in classifying audio waves.

In the following section, we will develop a time series forecasting model using 1D convolutions.

1D convolution for time series forecasting

We will continue to use the air pollution dataset to demonstrate 1D convolution for time series forecasting. The shape of input to the convolution layer is `(number of samples, number of timesteps, number of features per timestep)`. In this case, `number of timesteps` is seven and `number of feature per timestep` is one as we are concerned about only air pressure, which is univariate. To develop a convolutional neural network, we need to import three new classes:

```
from keras.layers.convolutional import ZeroPadding1D
from keras.layers.convolutional import Conv1D
from keras.layers.pooling import AveragePooling1D
```

A `ZeroPadding1D` layer is added after the input layer to add zeros at the beginning and end of each series. Zero-padding ensures that the convolution layer does not reduce the dimension of the output sequences. Pooling layer, added after the convolution layer, is used to downsample the input:

```
zeropadding_layer = ZeroPadding1D(padding=1)(input_layer)
```

Next, we add a `Conv1D` layer. The first argument of `Conv1D` is the number of filters, which determine the number of features in the output. The second argument indicates the length of the 1D convolution window. The third argument is `strides` and represents the number of places to shift the convolution window. Lastly, setting `use_bias` as `True` adds a bias value during the computation of an output feature. Here, the 1D convolution can be thought of as generating local AR models over a rolling window of three time units.

```
conv1D_layer = Conv1D(64, 3, strides=1, use_bias=True)(zeropadding_layer)
```

`AveragePooling1D` is added next to downsample the input by taking average over three timesteps with stride of one timestep. The average pooling in this case can be thought of as taking moving averages over a rolling window of three time units. We have used average pooling instead of max pooling to generate the moving averages:

```
avgpooling_layer = AveragePooling1D(pool_size=3, strides=1)(conv1D_layer)
```

The preceding pooling layer returns a 3D output. Hence, before passing to the output layer, a `Flatten` layer is added. The `Flatten` layer reshapes the input to `(number of samples, number of timesteps×number of features per timestep)`, which is then fed to the output layer after passing through a `Dropout` layer:

```
flatten_layer = Flatten()(avgpooling_layer)
dropout_layer = Dropout(0.2)(flatten_layer)
output_layer = Dense(1, activation='linear')(dropout_layer)
```

The layers are packed into a `keras.models.Model` wrapper and trained to minimize MAE using the Adam optimizer. Validation R-squared of the best model is 0.9933.

The detailed implementation is in the Jupyter notebook, `code/Chapter_5_Air_Pressure_Time_Series_Forecasting_by_1D_Convolution.ipynb`. We have taken a similar approach to develop a 1D CNN-based time series forecasting model for `pm2.5` and the code is written in the Jupyter notebook, `code/Chapter_5_PM2.5_Time_Series_Forecasting_by_1D_Convolution.ipynb`.

Summary

In this chapter, we have described three deep learning-based approaches to develop time series forecasting models. Neural networks are suitable in cases where there is little information about the underlying properties such as long-term trend and seasonality or these are too complex to be modeled with an acceptable degree of accuracy by traditional statistical methods. Different neural network architectures such as MLP, RNN, and CNN extract complex patterns from the data. If neural network models are trained with appropriate measures to avoid overfitting on training data, then these models generalize well on unseen validation or test data. To avoid overfitting, we applied dropout, which is widely used in deep neural networks for a variety of datasets and applications. We hope that this chapter gives you an idea of advanced techniques available for time series forecasting. The Jupyter notebooks accompanying the chapter are expected to give you the necessary base knowledge, which would be useful to develop more advanced models for complex problems.

Getting Started with Python

As you have chosen to read this book, we think that you might have a working knowledge of Python-if not, a hands-on expert who happens to live and breathe Python. In case you have a fair knowledge of Python at the least, you may choose to skip this appendix. If you are new to Python or looking for how to get started with the programming language, reading this appendix will help you get through the initial hurdles. It would also get you what you need to enjoy this book's chapters. So, without further ado, let's jump in!

Python is a general-purpose, high-level, and interpreted programming language, which appeared in 1991. Its creator, Guido van Rossum, started writing the interpretation of the language over the Christmas of 1989 and named the language after one of his favorite TV shows-Monty Python's Flying Circus.

Python emphasizes code readability through whitespace indentation to delimit code blocks rather than curly brackets, which are famously used in C, C++, and Java. Another powerful feature of Python is its succinctness, which allows programmers to express concepts in lines of code fewer than in C, C++, and Java. For example, the use of lambda functions as a quick and effective way of declaring functions in just one line is a favorite among Python developers. (Don't worry; we will get into what a lambda function is later.) Other than these, Python supports dynamic type binding and automatically handles memory for you. Besides, Python interpreters are available for all major operating systems. CPython is the most popular open source implementation of the interpreter. Moreover, Python is supported by hundreds of packages of wide-ranging functionalities such as web development, GUI building, advanced memory management, file handling for diverse file formats, scientific and numeric computing, image processing, machine learning, deep learning, big data, and many others. **PyPI** is the official repository of Python packages and almost all well-known packages are available there. The link to PyPI's website is `https://pypi.python.org/pypi`. However, we do not have to download packages from this website before installing. There are **command line** (**CL**) tools that do the job for us. These tools are available once a basic version of the language is installed. Moreover, these CL tools take a lot of work from us by ensuring that the requested package is compatible with the current version of Python and that the dependencies are already installed or need to be installed.

This appendix covers the following topics:

- Installation
- Basic data types
- Keywords, control statements, and functions
- Iterators and generators
- Classes and objects

Installation

Setting up Python on your computer would require installation of three components-the interpreter, an **integrated development environment** (**IDE**), and packages to support application development. Thankfully, we do not have to install these separately. There are software packages that bundle all three and a single installation makes the most available for use.

Python installers

There are two widely used options of installers that bundle the interpreter, an IDE, and useful Python packages:

- **Option 1**: Installers that bundle all three components are available for download from `https://www.python.org`. This website hosts the installers for several different versions of the language. The latest versions, at the time of writing this book, are Python 3.2.6 and Python 2.7.13. The installers come with a basic IDE that is a good place to write some code and get a feel of how the language works. Several packages such as the ones on file handling, memory management, mathematical computations, and so on come out of the box from these installers. However, most of the packages on scientific computing, statistical modeling, and machine learning need to be separately installed along with their dependencies. This makes this option time-consuming.

- **Option 2**: We can download freely available installers from `https://www.continuum.io/downloads`. Continuum Analytics have done a great job in putting together the interpreter, an IDE, and hundreds of useful packages in the same installer. These packages cover most of the things that you might need in order to work on data science projects. Jupyter Notebook and Jupyter Lab, the popular web-based development tools for Python, also come in the bundle. The full list of packages is at `https://docs.continuum.io/anaconda/packages/pkg-docs`. These installers are known as Anaconda distributions of Python. Both Python 2.7 and 3.6 versions are available. Most of the packages used in this book's examples came as a part of the Anaconda distribution. We had to install only Keras and Tensorflow separately.

The next section describes the steps required for this book's examples.

Running the examples

The code samples of this book have been written using the Jupyter Notebook development environment. All code files have the.ipynb extension and can be found in the code folder of the book's GitHub repository.

To run the Jupyter Notebooks, you need to install Anaconda Python Distribution. We downloaded Python 3.6 version of the distribution from `https://www.continuum.io/downloads`.

After downloading the installer, the following steps need to be performed to set up the environment for the examples:

- Install the Anaconda Python Distribution. Follow the prompts of a standard installation. The installer adds the path to Python executables to the PATH environment variable. The installer gives a prompt to add to the PATH variable. Make sure to click **Yes** or **OK** after reading the instructions, which will allow the installer to update the PATH.
- Open a new command prompt window and run the following command:

```
python --version
```

- This is to ensure that Python executables have been added to the PATH variable.
- Now it's time to set up the programming environment to run the Jupyter Notebooks:
    - Clone or download the book's GitHub repository, Practical Time Series, to a local disk, for example, D.
    - Open a new command prompt and run the following commands:

```
D: (optional if default location is some other drive)
cd "Practical Time Series"
jupyter notebook -notebook-dir="."
```

- The third step starts the Jupyter Notebook server and automatically opens the home page in the default browser. The home page shows the folders and files of the disk folder, D:\Practical Time Series. Click on the code folder in the home page and you will see the Jupyter notebooks. Every notebook has the following line that sets the current working directory for the notebook:

```
os.chdir('D:/Practical Time Series')
```

- If you have saved the Practical Time Series folder in the D drive, then there is no need to change this line. Otherwise, this must be changed to reflect the actual current working directory. For example, if the shared folder was saved as C:\Local\Practical Time Series, then change the preceding line of code to the following:

```
os.chdir('C/Local/Practical Time Series')
```

- Code snippets in a Jupyter notebook appear in rectangular cells with a grey background, as shown in the following screenshot. To run the notebooks, execute each cell sequentially from the top by clicking on the 'Run Cell' button (the eigth button from the left on the buttons panel):

Basic data types

Python supports basic numeric data types such as int, long, and float, just like all major programming languages. None represents a null pointer in Python.

Strings are sequential data types in Python. Other sequential data types that are commonly used in Python are lists and tuples. The difference between a list and tuple is that the former is mutable while the latter is an immutable type. Therefore, the interpreter would throw an error if you try to modify a tuple. Let's dig a little deeper into lists and tuples.

List, tuple, and set

A list is a collection of elements. An element can be of any data type and a list can contain elements of different data types. A list has several important functions such as `append`, `extend`, `insert`, `index`, `pop`, and few others. The following table summarizes the functionality of these functions:

Function name	Functionality
`append`	Adds a new element to the end of the list.
`extend`	Adds new elements from an iterable. The members of the iterable element are entered separately into the list. This is demonstrated in the code sample after this table.
`insert`	Adds a new element but at a specific position in the list.

index	Returns the position of the argument in the list. If the argument is not found in the list, then ValueError is thrown.
pop	Removes the element from the position indicated by the function's argument. If this index is not found, then an IndexError is thrown.

A list has a few other functions as well. For a more comprehensive summary, we recommend any material that teaches the Python language.

The functions mentioned in the preceding table are demonstrated in the Jupyter Notebook, code/Getting_started.ipynb. Note that Python uses zero as the index of the first element of a sequence. Hence, mylist[0] is the first element, mylist[1] is the second, and so on.

Unlike lists, tuples are immutable sequences and do not allow any alteration once they are declared. Tuples support only two functions, namely, count and index, which return the number of elements in a tuple and the index of an existing element. Requesting the index of a non-existent element throws a ValueError.

The len function can be used to find the length of a list as well as a tuple by passing the variable as an argument to the function.

Besides, lists and tuples are iterables. The technical definition of iterable is that it is any object that returns an iterator. We will discuss iterables and iterators in detail in the following sections. However, for the time being, just keep in mind that iterators are objects whose individual elements can be assessed by running a loop for the object. The following code demonstrates iterating over a list and a tuple:

```
#iterating over a list
print('Elements in mylist:')
for i in mylist:
print(i)
print('Elements in mytuple:')
for i in mytuple:
print(i)
```

The output of the preceding code snippet is as follows:

```
Elements in mylist:
-1
2
3
4
5
6
```

```
[7, 8]
7
8
Elements in mytuple:
1
2
3
4
```

Elements in a list and tuple can also be sliced for a range of indices. Slicing is an effective way of creating a subset from the original sequence:

```
#slicing lists and tuple
print('Second to fifth elemnts of mylist:', mylist[1:6])
print('Last three elemnts of mytuple:', mytuple[-3:])
```

This returns the following output:

```
Second to fifth elements of mylist: [2, 3, 4, 5, 6]
Last three elements of mytuple: (2, 3, 4)
```

Notice the negative index used to slice `mytuple` in the preceding code. Negative indices allow traversing from the end. Hence, -3 indicates the third position from the end. The first position from the end is an index of -1.

Sets are ordered sequences of unique elements and support standard set operations such as unions, intersections, and differences. Sets are created by passing an iterable. Set supports the pop operation but unlike list, the pop function removes only the first member of the set. Examples of sets are given in the Jupyter Notebook, `code/Getting_started.ipynb`.

Strings

Strings are iterables of 8-bit characters in Python 2.x and Unicode characters in Python 3.x. Strings can be declared in multiple ways, some of which are shown in the following code snippet:

```
#string declarations
mystring = 'Practical Time Series'
print(mystring)
mystring = "Practical Time Series"
print(mystring)
mystring = """Practical Time Series"""
print(mystring)
mystring = '''Practical Time Series'''
print(mystring)
```

All four of the preceding declarations result in the same string and the `print` function in each case gives the same output:

```
Practical Time Series
```

Strings are objects of type class, str, and support several functions that make string handling easy. We have shown a few examples in the Jupyter Notebook, `code/Getting_started.ipynb`.

Maps

Python supports hashmaps or associative arrays. Every element in a hashmap is associated with a definition or key. Dict is the basic hashmap implementation in Python. The elements in a dict are called values, which are mapped to references known as keys. A dict can be instantiated from an iterable of key-value pairs. This iterable can be a list, tuple, or an iterator, which we will discuss in one of the upcoming sections. The following code snippet shows multiple ways to create a dict:

```
mydict = dict([(0,1), (1,2), (2,3), (3,4), (4,5)])
mydict = dict([[0,1], [1,2], [2,3], [3,4], [4,5]])
mydict = {0:1, 1:2, 2:3, 3:4, 4:5}
```

Elements from a dict can be retrieved by invoking the get method that takes the key as input. Another way is to use `dictionary_variable[key]`. For non-existent keys, the get function returns None but the second method throws a `KeyError` error. The function item returns an iterator of key-value pairs. All pairs are not returned in a list but rather can be looped over the iterator.

Python is an object-oriented programming language and hence supports complex data types such as objects, which are instances of a class and have attributes and methods. Classes and objects will be covered in one of the following sections.

Keywords and functions

Keywords are reserved words that cannot be used as variable names. The following table gives the list of keywords and their purpose:

Keywords	Explanation
False	Boolean false value: `>>bool_var = False`

True	Boolean true value: `>>bool_var = True`

and	Logical operator that returns True only if both the operands are True or evaluates to True:
	`>>a = 2`
	`>>if a > 0:`
	`print(a)`
	`>>2`

as	Creates an alias for a module that is being imported:
	`import pandas as pd`
	`import scikit-learn as skl`

assert	Used to evaluate a logical expression to check values of variables at runtime and raises an AssertionError if the expression evaluates to False:
	`>>a = -2`
	`>> assert a > 0`
	The preceding assert keyword raises an AssertionError.

break	Used to exit a loop such as for loop or while loop when a condition is met.
class	Keyword that indicates a class declaration.
continue	Indicates the interpreter to move to the next iteration in a `for` or `while` loop without executing the code in the loop that follows the continue statement.
def	Indicates a function declaration.
del	Used to delete a variable and free the memory associated with it.
if	Used to check whether the following variable or expression meets a condition and executes the code if the condition is satisfied.
else	Executes the code if a Boolean condition that is checked by a preceding if statement is not met.
elif	Checks a Boolean condition when the preceding condition in the if statement is not met.
try	Declares scope for exception handling and passes the control to the scope within the following except keyword upon occurrence of an error.
except	Defines the scope and the code in it that will be executed when the preceding try block throws an error.

`finally`	Defines the scope and the code that needs to be always executed, regardless of exception generation from code, which is declared in the preceding try keyword.
`for`	Declares the scope for an iterative loop that iterates over the elements of a sequence.
`while`	Declares the scope for an iterative loop that runs until the condition in the `while` statement is satisfied.

`in`	Used to check whether values are present in a sequence such as a list or tuple:
	`>>mylist = [1, 2, 3]`
	`>>var = 1`
	`>>var in mylist`
	`>>True`
	Also used to traverse a sequence:
	`>>for i in mylist:`
	`print(i)`
	`>>1`

`import`	Used to import a package:
	`import os`
	`import sys`

`is`	Tests whether two variables refer to the same object or not. The == operator checks whether two variables are equal are not.

`lambda`	Used to create an inline function:
	`>>func = lambda x: x+1`
	`>>func(2)`
	`>>3`

`with`	Wraps the execution of a block of code within methods defined by the context manager, which is a class that implements the __enter__ and __exit__ methods. The __exit__ method is called at the end of the nested block of code:
	`with open('foo.txt', 'w') as f:`
	`f.write('Practical Time Series')`
	After this line is executed, the file is closed. This is possible because file objects have __enter__ and __exit__ methods.

`yield`	Returns a generator, which is covered in one of the following sections.
`return`	Exits a function and returns a value from it.

raise	Used to explicitly raise an error: >>a = -2 >>if a < 0: raise ValueError The preceding lines of code generate a ValueError.

In Python, functions are of three types:

- Inline functions
- Normal functions
- Built-in functions

Inline functions are declared using the lambda keyword. This type of function declaration can be named or anonymous. Anonymous inline functions are used to declare a function within another function or code block that loops over a sequence and applies the inline function to its elements. Named inline functions are one-line function declarations but have a function name:

```
#Inline anonymous function
mylist = [0, 1, 2, 3, 4, 5]
processed_list = list(map(lambda x: x+1, mylist))

#Inline named function
myfunc = lambda x: x+1
processed_list = [myfunc(i) for i in mylist]
print(processed_list)
```

Notice the use of the map function in the preceding code block. map is one of the predefined functions in Python. It takes an anonymous inline function and an iterable as input and applies the inline function to each element of the iterables. The output of a map function is a new iterable that has values returned by the inline function for each element of the input sequence but transformed by the inline function.

Normal functions are defined with the def keyword. Without a return statement, normal functions execute the code within their scope and return None. With a return statement, the output of a normal function can be obtained outside the function at the point of its invocation:

```
#Normal function which returns None
def myfunc(a):
print(Within the function:', a+1)
print('Return value of the function:', myfunc(5))
```

The output of the print function is as follows:

```
Within the function: 6
Return value of the function: None
```

```
#Normal function which returns an output
def myfunc(a):
a = a+1
print('Wihtin the function:', a)
return a
print('Return value of the function:', myfunc(5))
```

The output of the last print function is as follows:

```
Within the function: 6
Return value of the function: 6
```

Lastly, Python has several predefined functions such as range, map, filter, zip, enumerate, and so on. Expressiveness and brevity are two of the key elements of Python's design philosophy. To this end, predefined functions help us achieve a lot more with just a few lines of code. The following table describes the functionality of the predefined functions:

Function	Use
range	Returns a sequence of integers from start to stop. In Python 2.x, range gives a list, but in python 3.x, it returns an iterator.
map	Iterates over a sequence and applies a function to transform every element. Both the function and sequence are given as input to map.
filter	Takes a sequence as an input and returns another sequence, which consists of elements that pass a condition. Both the function and sequence are passed as input to filter.
zip	Takes two or more sequences as input and generates a sequence of tuples, each of which consists of elements from the input sequences.
enumerate	Generates (index, value) pairs from the element of a sequence.

In Python 2.x, all these functions return a list as output but in Python 3.x, they return an iterator. To get the actual output, one must loop over the iterator. This design is useful in case the input sequence is huge and too big to fit in the computer's memory. For example, consider a huge text file that must be read line by line along with line numbers. In Python, files can be read line by line by looping over an iterator returned by the file reader. To obtain line numbers along with the lines, we can invoke enumerate with the file iterator. The enumerate function treats the file iterator as a sequence and generates an iterator of (`line_number`, line) pairs, which can be looped one by one without loading the entire file into memory. A similar approach can be taken with the other built-in functions, which we have just mentioned. In the Jupyter Notebook, `code/Getting_started.ipynb`, we have shown how these functions can be used along with a file reader. We encourage you to try similar approach with other types of sequences.

We have covered some of the most frequently used built-in functions. However, it is recommended that you refer to a book or tutorial on the Python programming language to get an exhaustive list.

Iterators, iterables, and generators

In Python, we frequently encounter iterators, iterables, and generators as these are efficient ways of looping over a data type or data structure that is a sequence or a sequence can be created out of it. One clear advantage of using these looping techniques is that they require less memory. So, when you must access a sequence element by element, these techniques become very useful because a large sequence does not need to be loaded into memory all at once. For example, if you need to find the square of the first one trillion positive integers, there is no need to create a data structure to hold all numbers in memory at the same time. Iterators, iterables, and generators can be used to generate and process these numbers sequentially. Another example is processing a large text file. The entire file might not fit in memory. Hence, if we need to process the file, for example, to find word count per line of the file, we can iteratively loop over the lines and process them one by one. As iterators, iterables, and generators are so useful, let's understand what they are and how to use them.

Iterators

Objects that are instances of classes that have __iter__ and next functions and can be used with a for loop to go over a sequence element by element are iterators. The __iter__ function makes an object recognizable as an iterator. The next function is invoked to get the elements of a sequence one by one. Every time the next function is called, it returns an element from the predefined sequence or it creates an element. Hence, the next function can implement the logic to create the elements. When there are no more elements, the __next__ function throws a StopIteration error.

Let's go through examples to understand how iterators work. We create an iterator to return elements from a predefined sequence:

```
class MyIterator(object):

def __init__(self, seq):
self.seq = seq
self.i = 0

def __iter__(self):
return self

def next(self):
if self.i < len(self.seq):
i = self.i
self.i += 1
return self.seq[i]
else:
raise StopIteration()
```

An object of the MyIterator class is created and the next function is called five times. The first four calls return an integer from the sequence but the last call causes the next function to throw a StopIteration error as the entire length of the sequence has been traversed by now:

```
itr = MyIterator([1,2,3,4])
print(itr.next())
print(itr.next())
print(itr.next())
print(itr.next())
print(itr.next())
```

Now let's declare an iterator that implements the data generation logic in the next function instead of returning elements from a predefined sequence:

```
class MyIterator(object):

def __init__(self, n):
self.n = n
self.count = 0

def __iter__(self):
return self

def next(self):
if self.count < self.n:
i = self.count
self.count += 1
return i
else:
raise StopIteration()
```

An object of the `MyIterator` class is created and the next function is called six times. The first five invocations return the positive integers generated by the next function but the last call throws a StopIteration error:

```
itr = MyIterator(5)
print(itr.next())
print(itr.next())
print(itr.next())
print(itr.next())
print(itr.next())
print(itr.next())
```

Iterables

Objects that do not have the __iter__ or next function but can be used to create an iterator are iterables. The built-in __iter__ function takes a sequential object as input and returns an iterator. Then, another built-in function, next, takes the iterator and returns the iterator's elements in every invocation. Lists and tuples are not iterators but can be used to create iterators using the next function. The following code snippet demonstrates using a list as an iterator:

```
mylist = [1,2,3,4]
mylist_iter = iter(mylist)
print(type(mylist))
print(type(mylist_iter))
```

```
<class 'list'>
<class 'list_iterator'>
```

Notice that `mylist` is of type list while `mylist_iter`, which is created by the iter function, is of type `list_iterator`.

Generators

Generators give the functionality of iterators but without having to write an entire class. In many cases, the logic of creating the elements of a sequence are implemented in the next function of an iterator. For example, the code to read a text file line by line and return the occurrence of a specific word in each line needs to be written in the next function of an iterator. Other functions such as __init__ and __iter__ are implemented in the class merely to suffice the requirements of developing an iterator.

Generators are special functions that simplify development. The yield keyword in a function signals the Python interpreter that the function is a generator. Using generator functions, we implement only the logic of creating the elements of the sequence in the generator function in a while loop. The yield function returns the elements each time the function is called.

Let's implement a generator function that returns a whole number every time it is called:

```
def int_gen():
count = 0
while True:
i = count
count += 1
yield i
```

We will assign the generator function to a variable and invoke the built-in next function on this variable. The next function is run five times the variable as an argument:

```
ig = int_gen()
print(next(ig))
print(next(ig))
print(next(ig))
print(next(ig))
print(next(ig))
```

The output of the print is as follows:

```
0
1
2
3
4
```

Every time the next function is called on ig, it returns the current value of the count variable and increments it by one. The count variable is maintained in the internal state of the generator and hence can return its latest value.

Classes and objects

A class is a logical grouping of variables and functions. The class keyword is used in Python to define such logical groupings. A class often represents a real-life entity, for example, book, author, publishers, and so on. Entities have properties, which are represented by the variables defined in a class. Functions in a class, often referred to as methods, define how data about an instance of the entity can be captured and transformed. An instance of a class is a single realization of the entity. For example, book is an entity whereas Practical Time Series Analysis is an instance of book. To create instances, we initiate an object of a class. Object definition involves assigning values to the variables of the class through the constructor function. This job is done by the __init__ method that takes input and assigns them to class variables. The __init__ method can internally call other functions based on the logic of creating the object. Let's define a class about books:

```python
import datetime
class Book(object):

    def __init__(self, name, date_of_publication, nb_pages, publisher):
        self.name = name
        self.date_of_publication = datetime.datetime.strptime(date_of_publication,
        '%Y-%m-%d')
        self.nb_pages = nb_pages
        self.publisher = publisher
        self.authors = []

    def add_author(self, author_name):
        self.authors.append(author_name)

    def print_date_of_publication(self, print_format='%Y-%m-%d'):
        print(self.date_of_publication.strftime('%d-%m-%Y'))
```

Now we will create an object of the `Book` class:

```
mybook = Book('Practical Time Series Analysis', '2017-09-15', 200, 'Packt')
```

Note that the __init__ method does not specify the authors of the book but creates an empty list, authors. We can invoke the add_author function on the `mybook` object to add authors. Additionally, `date_of_publication` is initially set in the `%Y-%m-%d` format. The `print_date_of_publication` function takes print_format as input and displays the date in another format. For example, we can print the publication date as `%d-%m-%Y`:

```
mybook.print_date_of_publication(print_format='%d-%m-%Y')
```

Summary

This appendix covers the basics of the Python programming language. Topics such as data types, keywords, functions, classes, iterators, iterables, and generators have been discussed. These programming techniques form the building blocks of Python.

This book's chapters use several concepts and programming techniques that have been discussed here.

Index